Two Manuals Combined into One Book

40-MM Grenade Launcher M203

And

MK 19, 40-MM Grenade Machine Gun

Department of Defense

Pentagon
Publishing

FM 3-22.31 (FM 23-31)
C1

Change 1
Field Manual
No. 3-22.31

Headquarters
Department of the Army
Washington, DC, 19 March 2007

40-mm Grenade Launcher, M203

1. Change FM 3-22.31, 13 February, 2003, as follows:

Remove old pages:	**Insert new pages:**
i through iv	I through iv
1-1 through 1-2	1-1 through 1-2
Chapter 2	Chapter 2
Chapter 3	Chapter 3
Chapter 4	Chapter 4
Chapter 5	Chapter 5
6-9 through 6-10	6-9 through 6-10
Glossary	Glossary
References	References
Index	Index
DA Form 2946-R	DA Form 2946-R

2. A star (*) marks new or changed material.

3. File this transmittal sheet in front of the publication.

DISTRIBUTION RESTRICTION: Approved for public release; distribution is unlimited.

By Order of the Secretary of the Army:

PETER J. SCHOOMAKER
General, United States Army
Chief of Staff

Official:

Joyce E. Morrow

JOYCE E. MORROW
Administrative Assistant to the
Secretary of the Army
0705702

DISTRIBUTION:
Regular Army, USAR, and ARNG: To be distributed in accordance with initial distribution number 115893, requirements for FM 3-22.31.

This page intentionally left blank.

FIELD MANUAL
NO. 3-22.31 (23-31)

HEADQUARTERS
DEPARTMENT OF THE ARMY
WASHINGTON, D.C., 19 March 2007

40-MM GRENADE LAUNCHER, M203

CONTENTS

DISTRIBUTION RESTRICTION: Approved for public release; distribution is unlimited.

*This publication supersedes FM 23-31, 20 September 1994.

PREFACE

This publication provides technical information on and training and combat techniques for the M203 grenade launcher. Intended users include leaders and designated grenadiers, who will use this information to successfully integrate the M203 into their combat operations. This manual discusses gunnery training and train-the-trainer, and it includes an appendix on the 40-mm grenade launcher, M79.

*This publication applies to the Regular Army, the Army National Guard (ARNG)/Army National Guard of the United States (ARNGUS), and the United States Army Reserve (USAR) unless otherwise stated.

*The tactical positions shown in this manual were drawn to enhance the reader's understanding of related subject material and are *not* tactically correct. Also, to avoid hiding important graphic information, uniforms may be shown plain rather than camouflaged.

*This publication prescribes DA Form 2946-R, *40-MM Grenade Launcher Scorecard.*

*The proponent for this publication is the US Army Training and Doctrine Command. The preparing agency is the US Army Infantry School. You may send comments and recommendations by any means, US mail, e-mail, fax, or telephone, as long as you use or follow the format of DA Form 2028, *Recommended Changes to Publications and Blank Forms*. You may also phone for more information:

E-mail	lu.lusano@us.army.mil
Phone	COM (706) 545-8623 or DSN 835-8623
Fax	COM (706) 545-8600 or DSN 835-8600
US Mail	Cmdt, USAIS ATTN: ATSH-INB 6650 Wilkin Drive , Bldg 74, Rm 102 Fort Benning, GA 31905-5593

*Unless this publication states otherwise, masculine nouns and pronouns may refer to either men or women.

CHAPTER 1
INTRODUCTION

This chapter discusses the training strategy and combat conditions for the 40-mm grenade launcher, M203. (Appendix A discusses the M79 model.)

* 1-1. TRAINING STRATEGY

An effective overall training strategy produces well-trained grenadiers and trainers by integrating resources into an effective year-round training program. Beginning with IET and continuing both in other institutions (NCOES, OBC, and CCC) and in the unit, such a program trains and sustains the individual and collective skills needed to perform the wartime mission. Specific training strategies are implemented by institutional and unit training programs, and supporting training strategies are implemented through the use of other resources such as publications, ranges, ammunition, training aids, devices, simulators, and simulations. The year-round program includes periodic preliminary marksmanship training followed by zeroing and range qualification firing. Other key elements of the program are training for the trainers and refresher training for nonfiring skills. The example in Figure 1-1 on page 1-3 shows the flow of unit sustainment training.

*a. **Institutional Training**. Training strategy begins with combat arms initial-entry training (IET), which trains soldiers in the standards of M203 gunnery tasks. Soldiers graduate with basic and advanced M203 skills that include maintaining the M203 and using it to hit a variety of targets. Other institutional training programs, such as NCOES, OBC, and CCC, reinforce these skills. Related soldier skills are integrated into tactical training (STP 21-1-SMCT).

b. **Unit Training**. Training continues in units, where, in addition to sustaining proficiency in skills gained in institutional training, leaders and soldiers develop and sustain new skills such as suppressive and supporting fire. These skills are integrated into collective training exercises to develop combat readiness. Preliminary marksmanship training is conducted before firings and as other opportunities arise. (Appendix B discusses an M203 unit training program.) To be effective, a unit training program focuses on three battlefield variables:

(1) ***Target***. Is the target moving or stationary, single or multiple?

(2) ***Grenadier***. Is the grenadier moving or stationary? Is he kneeling, prone, or standing?

(3) ***Conditions***. Is visibility full or limited? Must soldiers wear protective masks or not? Is it day or night?

c. **Initial and Sustainment Training**. A task that is taught correctly and learned well is retained longer, so initial training is critical. In addition to being more easily sustained, well-trained skills are also easier to regain if not used for some time. Retraining may be needed, however, if too much time elapses, if training doctrine changes, or if personnel turnover is high.

(1) ***Collective Training***. Collective training exercises progress from drills (squad, section, and platoon) to STXs, and then to live-fire tactical exercises (LFXs). Drill books

and MTPs provide tasks and guidance needed to plan and conduct the exercises. After each, leaders and trainers conduct an AAR to evaluate both individual and unit proficiency. The results provide readiness indicators and requirements for future training. LFXs provide leaders with an overview of unit proficiency and training effectiveness. They can be conducted on any range approved for M203 firing.

(2) ***Leader Training***. The most critical part of the Army's overall gunnery training strategy is to train the trainers and leaders first. Leader courses, however, include only limited M203 training, so officers and NCOs should use available publications to develop their proficiency with the M203. Publications help leaders plan, conduct, and evaluate their gunnery training programs. Proponent schools provide training support materials (field manuals, training aids, devices, simulators, and audiovisual programs), which provide the doctrinal foundations for training the force.

(3) ***Advanced Training***. Once the soldier knows the weapon and has demonstrated skill in zeroing, training strategy provides for additional live-fire training and target-acquisition exercises, which are conducted at various ranges. To develop proficiency, soldiers must master different types of targets and scenarios of increasing difficulty.

(4) ***Proficiency Assessment***. This is conducted on the zeroing and record live-fire exercise range when soldiers complete IET.

1-2. COMBAT CONDITIONS

The trainer must realize that qualification is not an end but a step toward reaching combat readiness. To reach combat readiness, the grenadier should consider his position, the capabilities of his weapon, and the following combat conditions:

a. Enemy personnel are seldom visible except when assaulting.

b. Most combat fire must be directed at an area where the enemy has been detected or is suspected but cannot be seen. Area targets consist of objects or outlines of men irregularly spaced along covered and concealed areas (ground folds, hedges, borders of woods).

c. Most combat targets can be detected by smoke, flash, dust, noise, or movement and are visible only for a moment.

d. Some combat targets can be engaged by using reference points, predetermined fire, or range card data.

e. The nature of the target and irregularities of terrain and vegetation may require a grenadier to use a variety of positions to place effective fire on the target. In a defensive situation, the grenadier usually fires from a supported fighting position.

f. Most combat targets have a low-contrast outline and are obscured. Therefore, choosing an aiming point in elevation is difficult.

g. Time-stressed fire in combat can be divided into three types:

- A single, fleeing target that must be engaged quickly.
- Area targets engaged within the time they remain available.
- A surprise target that must be engaged at once with instinctive, accurate fire.

*CHAPTER 2

DESCRIPTION AND MAINTENANCE

Proper weapon maintenance is a vital part of all gunnery training programs. Good maintenance contributes to weapon effectiveness as well as to unit readiness. This chapter provides a technical description of the M203 grenade launcher, its components, and its ammunition. It also discusses proper procedures for clearing, disassembling, cleaning, lubricating, inspecting, and caring for the weapon.

2-1. DESCRIPTION

The M203-series grenade launchers are lightweight, single-shot, breech-loaded, pump action (sliding barrel), shoulder-fired weapons (Figure 2-1).

a. The original model of the M203 was designed for use with M16-series rifles.

b. The M203A1 model was designed for use with M4 series carbines.

c. The M203A2 model has a quick-release bracket for use with M4 and M4A1 carbines that have the M4 rail system and with M16A4 rifles that have the M5 adapter rail system.

Model	Who Mounts/Dismounts	Weapon Mounted On
M203	Unit armorer	M16-series rifles
M203A1/A2	DS maintenance co	M4-series carbine rifles

NOTE: Individual soldiers should neither mount nor dismount the grenade launcher.

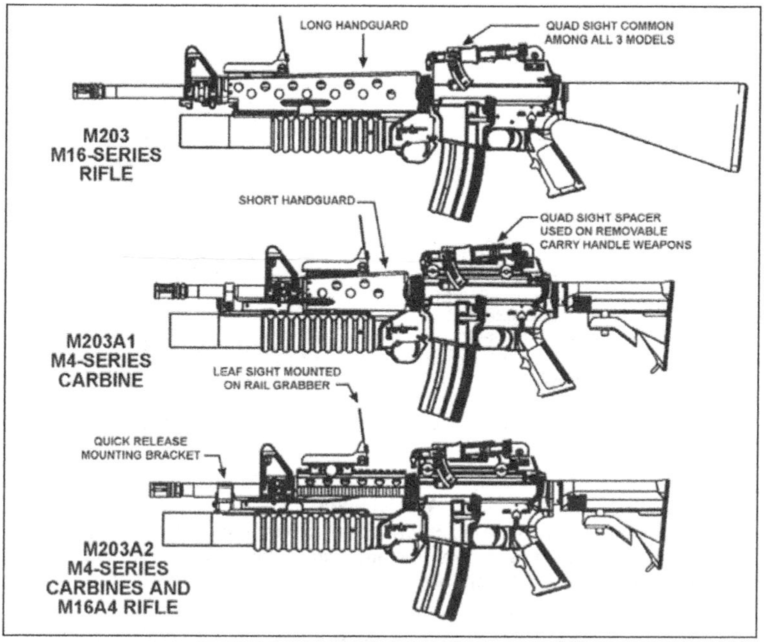

Figure 2-1. M203-series grenade launchers.

2-2. TECHNICAL DATA

Technical data for the M203-series grenade launchers follow:

a. **Weapon.**

Length
- Rifle and grenade launcher (overall) .... 99.0 cm (39 inches)
- Barrel only .... 30.5 cm (12 inches)
- Rifling .... 25.4 cm (10 inches)

Weight
- Launcher, unloaded .... 1.4 kg (3.0 pounds)
- Launcher, loaded .... 1.6 kg (3.5 pounds)
- Rifle and grenade launcher, both fully loaded .... 5.0 kg (11.0 pounds)

Number of lands .... 6 right hand twist

b. **Ammunition.**

Caliber .... 40 mm
Weight .... About 227 g (8 oz)

c. **Operational Characteristics.**

Action .... Single shot

Sights (use together or separately)	
Front leaf sight	Use between 50 and 250 meters (beyond 250 meters, the M16's barrel hides the target)
Rear quadrant sight	Use with M16's front sight post for ranges 50 to 250 meters.
Integrated front sight post, rear sight aperture	Use for 50- to 400 meters.
Chamber pressure	82,737 kilopascals (12,000 psi)
Muzzle velocity	76 mps (250 fps)
Maximum range	About 400 m (1,320 ft)
Maximum effective range	
Fire-team sized area target	350 m (1,155 ft)
Vehicle or weapon point target	150 m (495 ft)
Minimum safe firing range (HE)	
Training unprotected	165 m (541 ft)
Protected firing position	Less than 165 m (541 ft)
Combat	31 m (102 ft)
Minimum arming range	14 to 38 m (46 to 125 ft)
Rate of fire	5 to 7 rounds per minute
Minimum combat load	36 HE rounds

WARNING

When firing close-in, such as in urban areas, trenches, and other restrictive terrain, consider the minimum arming range to ensure the round clears other friendly forces.

2-3. COMPONENTS

Figure 2-2 shows the M203's major components, and the following paragraphs describe their purposes. Figure 2-3 through Figure 2-6 show the sight assemblies, the trigger and trigger guard, and the safety:

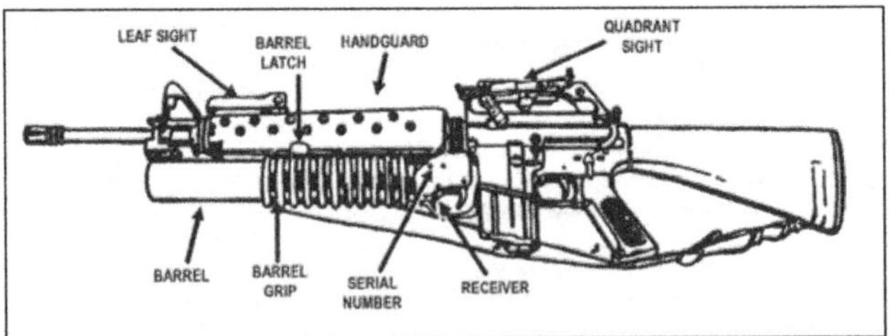

Figure 2-2. Components of the M203 grenade launcher.

a. **Handguard**. The handguard assembly houses the rifle barrel (Figure 2-2).

b. **Quadrant Sight Assembly**. The quadrant sight assembly, which attaches to the left side of the rifle's carrying handle, enables the grenadier to adjust for elevation and windage (Figure 2-2). This assembly consists of the sight, mounting screw, sight latch, rear sight aperture, sight aperture arm, front sight post, and sight post arm (Figure 2-3).

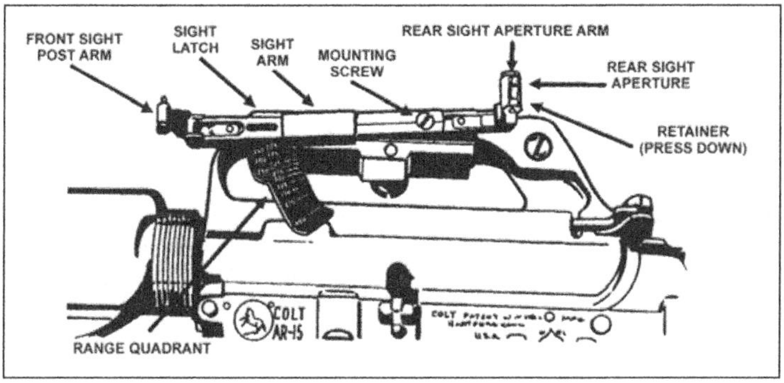

Figure 2-3. Quadrant sight assembly.

(1) ***Clamp, Bracket Assembly, and Mounting Screw***. The clamp and the bracket assembly hold the quadrant sight on the rifle's carrying handle. The mounting screw inserts through the right side of the clamp and into the bracket assembly.

(2) ***Sight Arm and Range Quadrant***. The sight arm mounts both the sight aperture arm (which holds the rear sight aperture) and the sight post arm (which holds the front sight post). This procedure allows the sight to pivot on the range quadrant to the desired range setting. The range quadrant is graduated in 25-meter increments from 50 to 400 m. Applying rearward pressure on the sight latch releases the quadrant sight arm so it can move along the range quadrant. Centering the number in the rear sight aperture selects the desired range. Releasing the sight latch locks the sight in position.

(3) ***Front Sight Post***. The front sight post mounts on the sight post arm by means of a pivot bracket. To prevent damage to the sights, keep the bracket closed when the sights are not in use. Use the sight post as follows to make minor adjustments in elevation when zeroing the launcher:

(a) To decrease elevation, turn the elevation adjustment screw on the sight post clockwise; to increase elevation, turn it counterclockwise.

(b) To move the impact of the projectile 5 meters at a range of 200 m, turn the elevation adjustment screw one full turn--360 degrees. To move the impact of the projectile 2.5 meters at a range of 200 m, turn the elevation adjustment screw one half turn--180 degrees.

(4) ***Rear Sight Aperture***. The rear sight aperture is on the sight aperture arm, which is attached to the rear portion of the quadrant sight arm. Use the rear sight aperture as follows to make minor adjustments in deflection (windage) when zeroing the launcher:

(a) To move the impact to the left, press the rear sight aperture retainer down and move the rear sight aperture away from the barrel; to move to the right, move it toward the barrel.

(b) To move the impact of the projectile 1.5 meter at a range of 200 m, move the rear sight aperture one notch.

c. **Receiver Assembly and Serial Number**. The receiver assembly houses the firing mechanism and ejection system and supports the barrel assembly. On the left side of the receiver assembly is the launcher's serial number (Figure 2-2).

d. **Barrel Assembly**. The barrel assembly holds the cartridges ready for firing and directs the projectile (Figure 2-2).

e. **Barrel Latch**. On the left side of the barrel is a latch that locks the barrel and receiver together (Figure 2-2). To open the barrel, depress the barrel latch and slide the barrel forward.

f. **Leaf Sight Assembly**. The leaf sight assembly is attached to the top of the handguard (Figure 2-4). The leaf sight assembly consists of the sight, its base and mount, an elevation adjustment screw, and a windage adjustment screw. Elevation and windage scales are marked on the mount. The folding, adjustable, open ladder design of the sight permits rapid firing without sight manipulation. The front sight post of the M16-series rifle serves as the front aiming post for the M203 leaf sight.

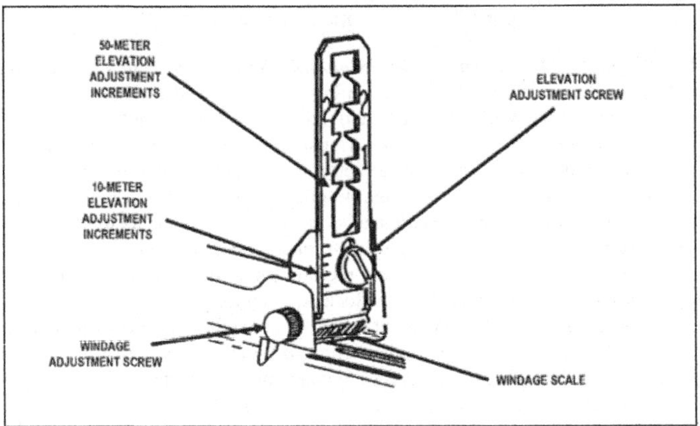

Figure 2-4. Leaf sight assembly.

(1) ***Sight Base***. Two mounting screws permanently attach the sight base to the rifle handguard. When the sight is down, the base protects it from damage.

(2) ***Sight Mount and Sight***. The grenadier uses the sight mount, which is attached to the sight base, to raise or lower the sight. Though the range is not marked on the sight in m, the sight is graduated in 50-meter increments from 50 to 250 m, which are marked with a "1" at 100 meters and a "2" at 200 m.

(3) ***Elevation Adjustment Screw and Elevation Scale***. The screw attaches the sight to its mount. When the screw is loosened, the sight can be moved up or down to make minor adjustments in elevation during the zeroing procedure. The rim of a 40-mm cartridge case is useful for turning the screw. Raising the sight increases the range;

lowering the sight decreases the range. The elevation scale consists of five lines spaced equally on the sight. The index line is to the left of the sight. Moving the sight one increment moves the impact of the projectile 10 meters in elevation at a range of 200 meters.

(4) ***Windage Screw and Windage Scale***. The firer uses the knob on the left end of the windage screw to make minor deflection adjustments during the zeroing. The scale has a zero line in the center. Two vertical lines are located on either side of and equidistant from the zero line. At a range of 200 meters, turning the knob on the windage scale one increment to the *left* moves the impact of the projectile 1.5 meter in the opposite direction, that is, to the *right*.

DANGER

NEVER **USE HIGH-EXPLOSIVE AMMO TO ZERO AT 50 METERS OR LESS--AT THAT RANGE, FRAGMENTATION COULD KILL YOU.**

TO REMIND YOU OF THIS, THE 50-METER MARK ON THE LEAF SIGHT BLADE IS PAINTED RED.

f. **Trigger Guard**. The trigger guard protects the trigger (A, Figure 2-5). Depressing the rear portion of the trigger guard rotates it down and away from the magazine well of the rifle. This gives the firer room to wear gloves or mittens while firing, if needed (B, Figure 2-5).

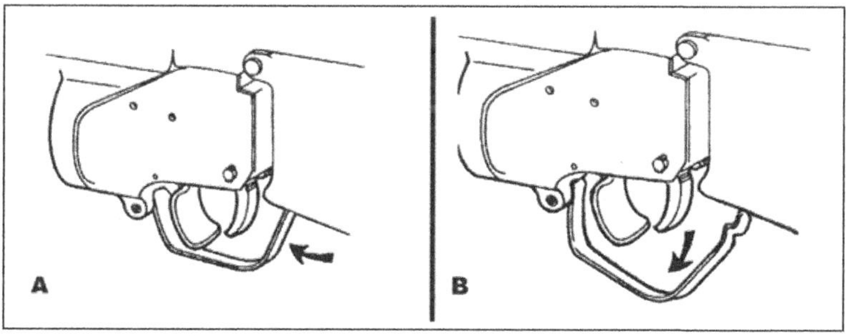

Figure 2-5. Trigger guard.

g. **Safety**. The safety is inside the trigger guard, just in front of the trigger. For the launcher to fire, the safety must be forward. When the safety is rearward, the launcher is on "S." The grenadier adjusts the safety manually (Figure 2-6).

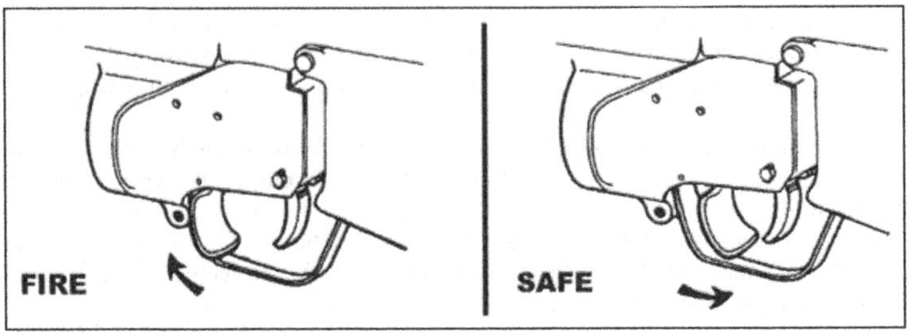

Figure 2-6. Safety.

2-4. AMMUNITION

The M203 grenade launcher uses several fixed-, low-velocity, 40-mm rounds. These include high-explosive, illuminating, signaling, CS, training, and multipurpose ammunition. This paragraph covers the most commonly used rounds.

WARNING

If fired into snow or mud, 40-mm rounds might hit too softly to detonate. An undetonated round might then explode when stepped on or driven over.

When training in snow or mud, avoid this hazard by firing only TP rounds.

a. **Types, Characteristics, and Capabilities**. Figure 2-7 shows and Table 2-1 compares the M203 rounds (TM 43-0001-28 provides more details.)

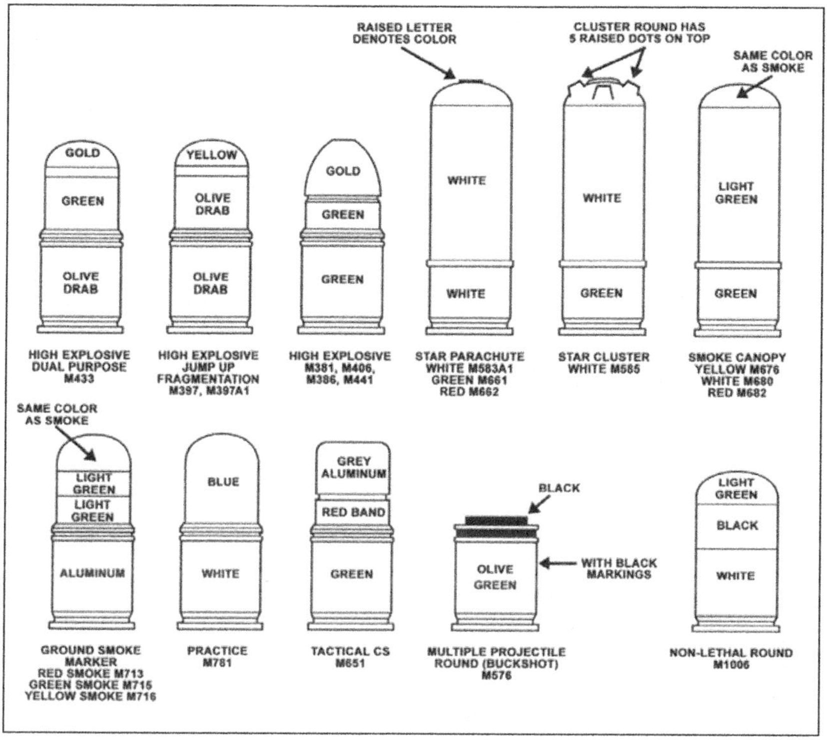

Figure 2-7. Cartridges for the M203 grenade launcher.

MODEL	DESCRIPTION	APPLICATION/EFFECTS
High Explosive, Dual-Purpose Round M433: DODAC 1310-B546 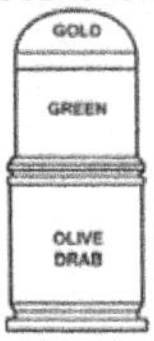 	Ogive: Gold-colored steel cup Body: Green Skirt: Olive drab aluminum Markings: White Length: 10.29 cm (4.05 in) Weight: 0.23 kg (0.51 lb)	This round has a small slug (1/4 inch diameter) that penetrates at least 5 cm (2 in) of steel armor when fired straight (flat, that is, at 0 degree obliquity) within 150 m. When fired at a point target, it arms between 14 and 27 m (46 to 89 ft). It can cause– • Casualties within 165 m. • Incapacitation within 15 m. • Death within 5 m.
	WARNING **In training, seek cover before firing an HE round at a target within 165 meters.** **Also in training, never engage targets within a 165 meter radius, that is, within 165 meters in any direction from unprotected friendly troops.**	
High-Explosive Jump-Up Fragmentation M397 and M397A1 DODAC 1310-B569 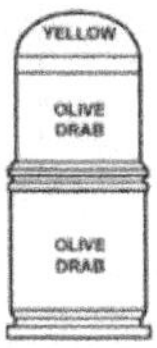 	Ogive: Yellow steel ball Skirt: Olive drab aluminum Filler: Explosive Markings: Yellow Length: 4.05 in. Weight: 0.51 lb	These rounds are designed to inflict personnel casualties using an air burst effect. It arms between 14 and 27 meters (46 and 89 feet). It detonates and fragments about 5 feet above the ground impact point.
High-Explosive M381: DODAC 1310-B568 M441: DODAC 1310-B575 M386: DODAC 1310-B574 M406: DODAC 1310-B568 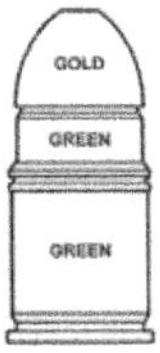 	Ogive: Gold Body: Green Skirt: Olive drab aluminum with attached steel projectile Markings: Gold Length: 3.89 in. Weight: M381: 5.03 lb; M441: 5.03 lb; M386: 0.50 lb; M406: 0.50 lb	These rounds produce a ground burst that causes casualties within a 165 meter radius. The kill radius is 5 meters. The M381 and M441 rounds arm within 2.4 to 3 meters (8 to 10 feet). The M386 and M406 rounds arm between 14 and 27 meters (46 to 89 feet).

Table 2-1. Comparison of M203 rounds.

MODEL	DESCRIPTION	APPLICATION/EFFECTS
Star Parachute Rounds M583A1 (WHITE) DODAC 1310-B531 90,000 CANDLEPOWER M661 (GREEN) DODAC 1310-B504 8,000 CANDLEPOWER M662 (RED) DODAC 1310-B505 20,000 CANDLEPOWER 	Cap: White, with raised letters indicating parachute color Body: White impact or bar alloy aluminum Skirt: Green Markings: Black Length: 13.39 cm (5.27 in) Weight: 0.22 kg (0.49 lb)	These rounds are used for illumination and signals, and are lighter and more accurate than comparable handheld signal rounds. The parachute attached to the round deploys upon ejection to lower the candle at 7 feet per second. The candle burns for about 40 seconds.
White Star Cluster M585 DODAC 1310-B536 	Cap: Plastic, with five raised dots for night identification. Body: White impact or bar aluminum alloy Skirt: Green Markings: Black Length: 13.38 cm (5.27 in) Weight: 0.19 kg (0.41 lb)	This round is used for illumination and signals. It is lighter and more accurate than comparable handheld signal rounds. The individual stars burn for about 7 seconds during free fall.
Ground Marker (Smoke) Rounds M713, (RED) DODAC 1310-B506 M715 (GREEN) DODAC 1310-B508 M716 (YELLOW) DODAC 1310-B509 	Ogive: Same as smoke color Body: Light green impact aluminum Skirt: Aluminum Markings: Black Length: 3.91 in. Weight: 0.49 lb	This round is used to mark soldiers' locations for others to see from the ground or the air. These rounds arm between 15 and 45 meters. If a fuze fails to function on impact, the output mixture in the front end of the delay casing backs up the impact feature.

Table 2-1. Comparison of M203 rounds (continued).

MODEL	DESCRIPTION		APPLICATION/EFFECTS
Smoke Canopy Rounds M676 (Yellow) DODAC 1310-B475 M680 (White) DODAC 1310-B477 M682 (RED) DODAC 1310-B479 	Ogive: Body: Skirt: Markings: Length: Weight:	Same as color of smoke Light green impact aluminum Aluminum Black 5.212 in. 0.48 lb	This round is used to mark soldiers' locations for others to see from the ground or the air. These rounds arm between 15 and 45 meters. Burst height is 91 meters.
Practice Round M781 DODAC 1310-B519 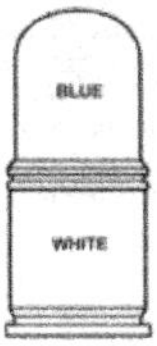	Windshield: Body: Skirt: Markings: Length: Weight:	Blue plastic Zinc or aluminum White White 10.29 cm (4.05 in) 0.22 kg (0.48 lb)	Round produces a yellow or orange signature on impact.
Tactical CS M651 DODAC 1310-B567 	Ogive: Band: Skirt: Markings: Length: Weight:	Grey aluminum Red Green Black markings 11.43 cm (4.50 in) 0.22 kg (0.48 lb)	This multipurpose round works best in riot control and urban operations. It arms between 10 to 30 meters (33 to 99 ft) and produces a white cloud of CS gas on impact.
Multiple Projectile (Buckshot) M576: DODAC 1310-B534 	Ogive: Body: Skirt: Markings: Length: Weight:	NA Olive drab NA Black 6.72 cm (2.65 in) 0.12 kg (0.25 lb)	This multipurpose round is best in thick, vegetated areas or for use in room clearing. It contains at least 20 pellets, which cast a cone of fire 30 meters wide and 30 meters high, and which travel at 269 meters per second. Buckshot rounds should be aimed at the foot of the target. This round has no mechanical type fuze.

Table 2-1. Comparison of M203 rounds (continued).

MODEL	DESCRIPTION	APPLICATION/EFFECTS
Nonlethal Round M1006: DODAC 1310-BA06 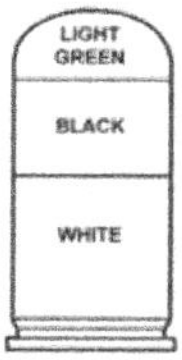	Nose: Green foam rubber Body: Black, high-density plastic Skirt: NA Other: Projectile assembly Cartridge case assembly Length: 3.95 in. Weight: 0.15 lb	This round is used to incapacitate personnel without penetrating the body. It is used to control riots, police, and keep the peace.

Table 2-1. Comparison of M203 rounds (continued).

b. **Storage**. Ammunition should be stored under cover. If this is not possible, it should be stored at least 15 centimeters (6 inches) aboveground, and covered with a double layer of tarpaulins. Tarpaulins should protect the ammunition, but allow ventilation. Trenches should be dug to prevent water from flowing under the ammunition.

c. **Care, Handling, and Preservation**. Ammunition containers should remain tightly closed until the ammunition is needed. Ammunition removed from the airtight container is likely to corrode, particularly in damp climates. Soldiers must also take the following precautions:

(1) ***Protect Ammunition from Mud, Dirt, and Water***. If the ammunition gets wet or dirty, wipe it off before use. Also, wipe off lightly corroded cartridges as soon as you find the corrosion. Never fire heavily corroded or dented projectiles or those with loose parts or particles.

> **WARNING**
> **Never fire heavily corroded or dented projectiles, or those with loose parts or particles.**

(2) ***Avoid Exposing Ammunition to the Direct Rays of the Sun.*** Hot powder can cause excessive pressure when the round is fired.

(3) ***Do Not Lubricate Ammunition***. This can cause dust and other abrasives to collect on it and damage the operating parts of the launcher.

d. **Packaging**. Ammunition packaging varies according to the type of ammunition:

(1) ***HE, HEDP, and TP***. Each wire-bound box of HE and HEDP ammunition contains 1 fiber box with 12 bandoleers of 6 rounds each, for a total of 72 rounds. Each wire-bound box of M781 practice rounds contains 4 fiber boxes of 25 rounds each, for a total of 100 rounds.

(2) ***Smoke and Cluster Ammunition***. Each wire-bound box of smoke and cluster ammunition contains 2 cans with 22 rounds each, for a total of 44 rounds.

(3) ***CS Ammunition***. Each box of CS ammunition contains 2 cans with 4 bandoleers of 6 rounds each, for a total of 48 rounds.

(4) ***Buckshot***. Each box of buckshot ammunition contains 12 bandoleers of 6 rounds each, for a total of 72 rounds.

(5) ***Nonlethal Ammunition.*** Each wire-bound wooden box contains 22 rounds per M2A1 metal box, for a total of 44 rounds.

3-5. CLEARING PROCEDURES

The soldier must clear the weapon before performing maintenance on it. FM 3-22.9 (23-9) provides instructions for clearing an M16-series rifle. To clear the grenade launcher--

a. Push in the release button and pull the barrel forward.

b. Watch to see if a round extracts.

c. Place the safety on "S."

d. Inspect the breech to ensure a round is not present.

e. Pull the barrel to the rear until it clicks. This cocks the weapon.

2--6. GENERAL DISASSEMBLY

As he disassembles the weapon, the soldier places each part on a clean, flat surface such as a table, shelter half, or disassembly mat. This aids in reassembly and simplifies the task of keeping up with the parts. The soldier will later reassemble the grenade launcher in the reverse order that he disassembled it. (Only ordnance personnel may disassemble the grenade launcher beyond the steps described here.) To disassemble the weapon--

a. Loosen the mounting screw and remove the quadrant sight assembly from the carrying handle of the M16-series rifle (Figure 2-8).

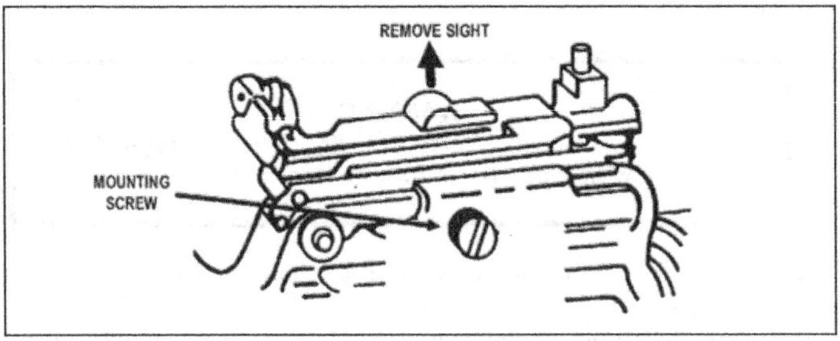

Figure 2-8. Removal of the quadrant sight assembly.

b. Remove the barrel assembly and handguard assembly, in either order:

(1) ***Barrel Assembly First.*** Push the barrel latch and move the barrel forward until it hits the barrel stop. On the left side of the handguard, insert a cleaning rod into the fourth hole back from the muzzle, depress the barrel stop, and slide the barrel forward and off (Figure 2-9).

Figure 2- 9. Removal of the barrel assembly before the handguard assembly.

(2) ***Handguard Assembly First***. Pull back on the M16's slip ring and remove the handguard by pulling it up and back. Push the barrel latch and move the barrel forward until it hits the barrel stop. Use a cleaning rod to depress the barrel stop and slide the barrel forward and off (Figure 2-10).

> **CAUTION**
> Never use a screwdriver or any other tool when removing the hand guard assembly. Doing so could damage the hand guard assembly or slip ring.

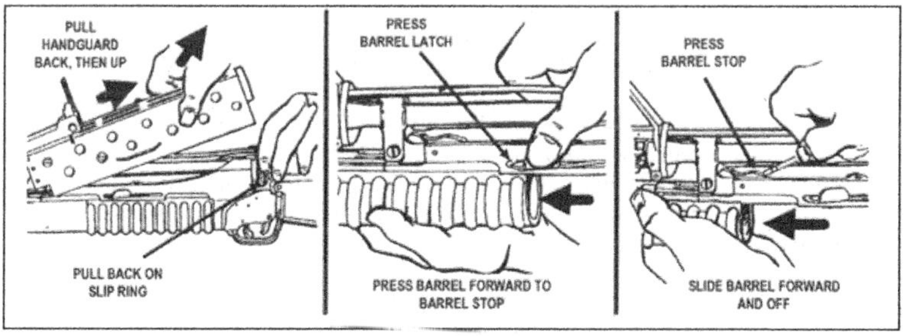

Figure 2-10. Removal of the handguard assembly before the barrel assembly.

2-7. CLEANING AND LUBRICATING PROCEDURES

After firing the grenade launcher, or if it has been idle for a long time, the soldier must clean and lubricate it as follows:

a. **Bore**. Attach a clean, dry rag to the thong, and then thoroughly moisten the rag with CLP. Pull the rag through the bore several times. Attach the bore brush to the thong, pull it through the bore several times, and follow this with more rags moistened with CLP (Figure 2-11). Pull dry rags through the bore, and inspect each rag as it is removed. The bore is clean when a dry rag comes out clean. Finally, pull a rag lightly moistened with CLP through the bore to leave a light coat of lubricant inside the barrel.

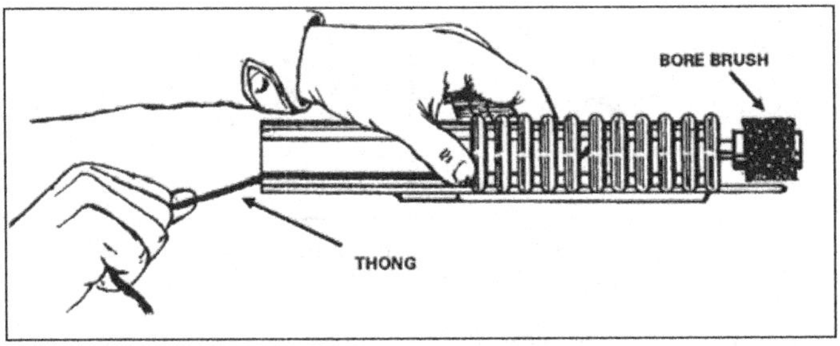

Figure 2-11. Cleaning of the barrel with thong and bore brush.

b. **Breech Insert**. Clean the face of the breech insert with a patch and CLP. Remove the CLP with dry rags; then lubricate the breech with a new, light coat of CLP.

c. **Other Parts**. Use a brush and dry rags to clean all the other parts and surfaces. After cleaning, apply a light coat of CLP to the outside of the launcher.

d. **Safety Mechanism**. Clean the safety mechanism properly with CLP; then lubricate it with CLP.

e. **Special Lubrication Requirements**. Lubricate the grenade launcher only with CLP and IAW the following environmental guidelines:

(1) ***Extreme Heat***. Lubricate with CLP, grade 2.

(2) ***Damp or Salty Air***. Clean the weapon and apply CLP, grade 2, frequently.

(3) ***Sandy or Dusty Air***. Clean the weapon and apply CLP, grade 2, frequently. Remove excess CLP with a rag after each application.

(4) ***Temperatures Below Freezing***. When the weapon is brought in from a cold area to a warm area, keep it wrapped in a parka or blanket, and allow it to reach room temperature gradually. If condensation forms on the weapon, dry and lubricate it at room temperature with CLP, grade 2, before returning it to cold weather. Otherwise, ice will form inside the mechanism.

NOTE: Although CLP provides sufficient lubrication down to -35°F (-37°C), it will only flow from its 1/2-ounce bottle, at 0°F (-17°C) and above.

2-8. INSPECTION

Inspection begins with the weapon already disassembled into its major groups or assemblies. Parts with shiny surfaces are serviceable. The following parts of the weapon and related equipment are inspected IAW TM 9-1010-221-10:

a. **All Parts**. Check for wear and damage, including burrs, scratches, and nicks.

b. **Handguard**. Check for cracks, dents, or distortion that prevents its firm attachment to the rifle.

c. **Leaf Sight Assembly**. Check for bent or damaged parts, rust or corrosion, and legible markings.

d. **Barrel**. Check for cracks or dents.

e. **Cartridge and Retainers**. Check for breakage, bends, chips, or missing parts.

NOTE: Take any unserviceable part to the armorer, who will decide if it is serviceable and replace parts as needed.

2-9. GENERAL ASSEMBLY

The soldier assembles the grenade launcher in the reverse order of disassembly.

a. Install the barrel by pressing the barrel stop and sliding the barrel into the receiver (Figure 2-12).

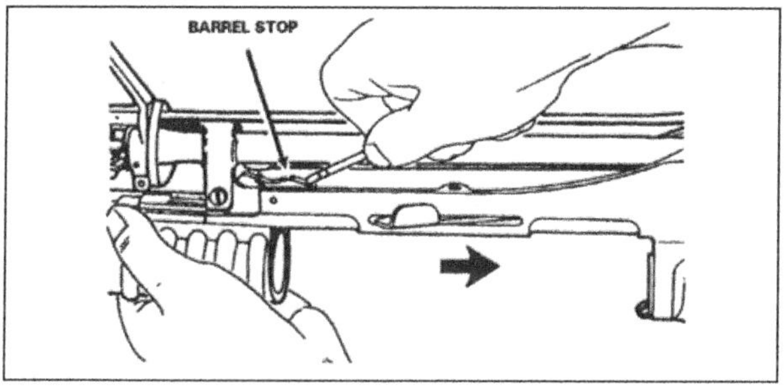

Figure 2-12. Installation of the barrel.

b. Lock the barrel by moving it rearward until it closes with a click (Figure 2-13).

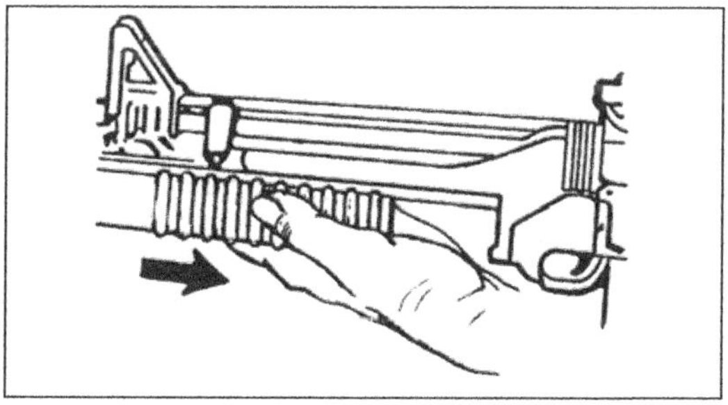

Figure 2-13. Locking of the barrel.

c. Install the handguard and secure it with the slip ring (Figure 2-14).

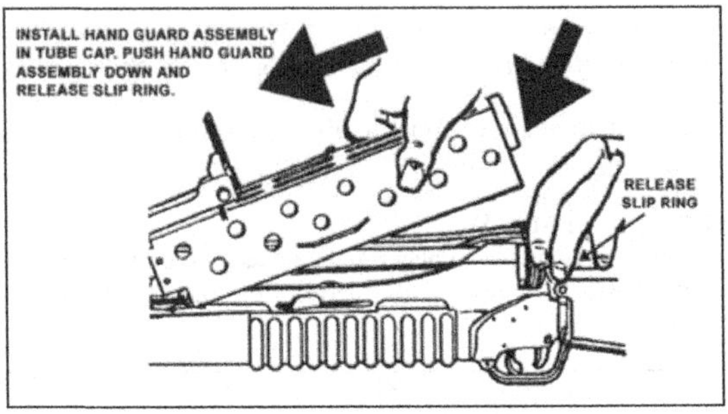

Figure 2-14. Installation and securing of the handguard.

d. Install the quadrant sight assembly (Figure 2-15).

Figure 2-15. Installation of the quadrant sight assembly.

e. Perform a function check to ensure that the grenade launcher has been assembled correctly. Notify the unit armorer at once if the launcher fails to function. Conduct the function check in this order:

(1) Check the proper operation of the sear. Cock the launcher, pull the trigger, and continue to hold the trigger. When you pull the trigger, you release the firing pin and should hear a metallic click The firing pin should release with a metallic click. Hold the trigger to the rear and cock the launcher again. Release the trigger, then pull. The firing pin should again release.

WARNING

If the sear malfunctions, the launcher could fire without your pulling the trigger.

(2) Check the safety by pulling the trigger in both the "S" and "F" positions. You must cock the launcher before you can place the safety in the "S" position.

(3) Check the windage adjustment screw on the leaf sight assembly for proper operation.

(4) Move the elevation adjustment screw only if the weapon has been zeroed.

(5) Move the barrel forward and back to be sure the barrel stop and barrel latch function.

2-10. CARE AND HANDLING PROCEDURES

Certain steps must be taken before, during, and after firing to properly maintain the grenade launcher.

a. **Before firing.**

- Wipe the bore dry.
- Inspect the weapon as outlined in the operator's technical manual.
- Ensure the weapon is properly lubricated.

b. **During firing.**

- Periodically inspect the weapon to ensure that it is lubricated.
- When malfunctions or stoppages occur, see the procedures in Chapter 4.

c. **During CBRN Conditions.** If contamination is anticipated, the soldier should apply CLP to all outer metal surfaces of the weapon. Ammunition, however, should never be lubricated. The soldier should keep the weapon covered as much as possible. If the weapon is contaminated, he should first decontaminate it IAW FM 3-3 and FM 3-11.4, and then he should clean and lubricate it.

2-11. DECONTAMINATION

Leaders must try to reduce the penetration of contaminants and exposure to them. Contaminated material is disposed of IAW SOP.

a. **Chemical.** Use towelettes from the M258A1 kit to wipe off the weapon. If these are unavailable, wash the weapon with hot soapy water.

b. **Biological.** Use towelettes from the M258A1 kit to wipe off the weapon. If these are unavailable, wash with soap and water.

c. **Radiological or Nuclear.** Wipe off the weapon with warm soapy water. Otherwise, use towelettes or rags (FM 3-11.5).

*CHAPTER 3

OPERATION AND FUNCTION

This chapter discusses the operation and function of the M203 grenade launcher.

3-1. OPERATION

The grenadier's operations include loading, unloading, and firing the weapon. The weapon uses a high-low propulsion system to fire a 40-mm round. The firing pin strikes the primer, whose flash ignites the propellant in the brass powder-charge cup inside the high-pressure chamber. The burning propellant produces 12,000 psi chamber pressure, which ruptures the brass powder-charge cup at the vent holes and allows the gases to escape to the low-pressure chamber in the cartridge case. There the pressure drops to 3,000 psi and propels the grenade from the muzzle at a velocity of 250 fps. The grenade's 3,700-rpm right-hand spin stabilizes it during flight and applies enough rotational force to arm the fuze. The weapon is unloaded with the barrel open and fired from a closed bolt. It must be cocked before it can be placed on safe.

3-2. LOADING OF M203 WITH THE SAFETY ON ("S")

When the safety is on "S," the grenadier loads the M203 as follows (Figure 3-1):

a. Keep the safety on "S" until ready to fire.

b. Depress the barrel latch.

c. Slide the barrel assembly forward. This automatically extracts a round and cocks the weapon.

CAUTION

Before loading the M203, make sure that the bore and chamber are clean and dry.

d. Insert clean, dry, undented ammunition into the chamber.

e. Slide the barrel rearward until it locks with an audible click.

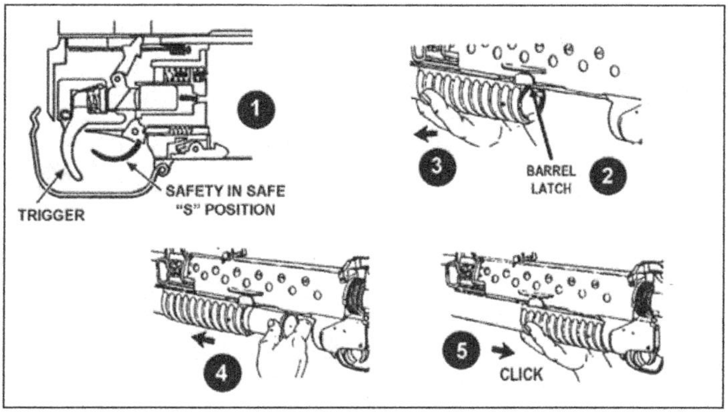

Figure 3-1. Loading the M203 with the safety on ("S").

3-3. LOADING OF M203 WITH THE SAFETY OFF ("F")

When the safety is off, the grenadier loads the M203 as follows (Figure 3-2):

a. Depress the barrel latch.

b. Slide the barrel assembly forward. This automatically extracts a round and cocks the weapon.

c. Place the weapon on safe ("S").

> **CAUTION**
> Before loading ensure bore and chamber is clean and dry.

d. Insert clean, dry, undented ammunition into the chamber.

e. Slide the barrel rearward until it locks with an audible click.

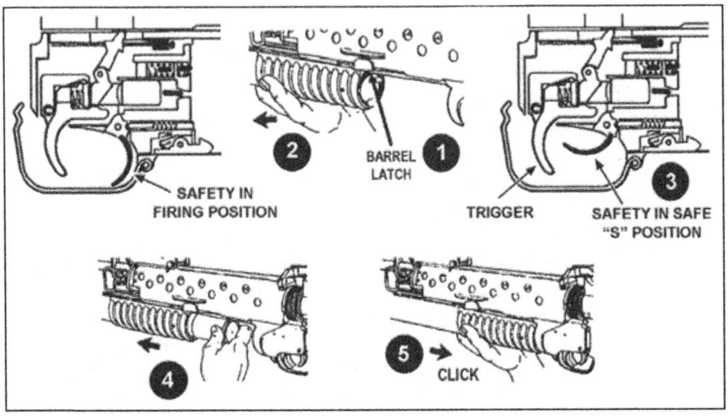

Figure 3-2. Loading the M203 with the safety off ("F").

> **WARNING**
>
> **Keep the muzzle pointed downrange and clear of all Soldiers. Use the correct ammunition:**
>
> - **Never fire pyrotechnic ammunition made for the AN-M8 pyrotechnic pistol in the M203.**
> - **Never load aircraft ammunition M384 (HE) or M385 practice.**
> - **Never use high-velocity 40-mm ammunition designated for other 40-mm weapons such as the MK 19. High-velocity rounds are longer than those used in the M203 and might cause this weapon to explode, resulting in serious injury.**

3-4. UNLOADING OF M203

To unload the grenade launcher, the grenadier must first depress the barrel latch and move the barrel forward. The cartridge case or round should automatically eject. If the case sticks, the grenadier taps it with a cleaning rod to remove it (Figure 3-3). He places the weapon on safe ("S"), then slides the barrel rearward, locking it to the breech.

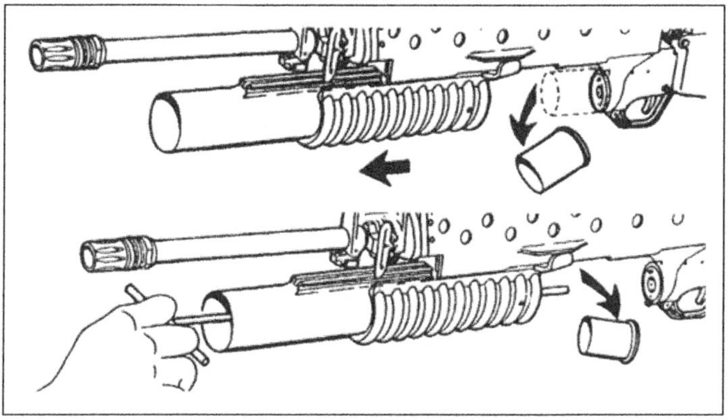

Figure 3-3. Unloading the M203 grenade launcher.

> **WARNING**
>
> **To unload an unfired weapon, avoid detonation by either--**
>
> - **Catching the ejected round or**
> - **Holding the weapon close to the ground to reduce the distance the round can fall.**

3-5. CYCLE OF FUNCTIONING

Knowing the M203's cycle of functioning from loading to firing helps grenadiers recognize and correct stoppages. Many of the actions described in this chapter occur at once, but here they are explained separately.

a. **Unlocking**. The cycle begins when the grenadier depresses the barrel latch to unlock the barrel assembly and slides the barrel assembly forward (Figure 3-4).

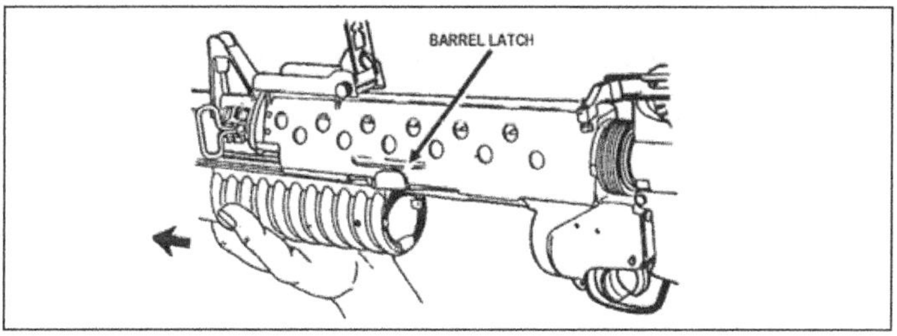

Figure 3-4. Unlocking the barrel assembly.

b. **Cocking**. The grenadier moves the barrel assembly forward, then backward, to cock the weapon. As the barrel assembly moves, it takes with it the barrel extension. Their movement causes the following to occur:

(1) The cocking lever is forced down as the barrel assembly and barrel extension, which are interlocked with the cocking lever, move forward.

(2) The movement of the cocking lever forces the spring-loaded firing pin to the rear.

(3) The spring-loaded follower also moves forward with the barrel extension.

(4) The barrel assembly continues forward, disengaging the barrel extension from the cocking lever. The cocking lever is then held down by the follower.

(5) When the grenadier begins to move the barrel assembly back to the rear, this forces the follower to the rear.

(6) The cocking lever again engages the barrel extension, which causes the firing pin to move slightly forward and engage the primary trigger sear. This cocks the weapon (Figure 3-5).

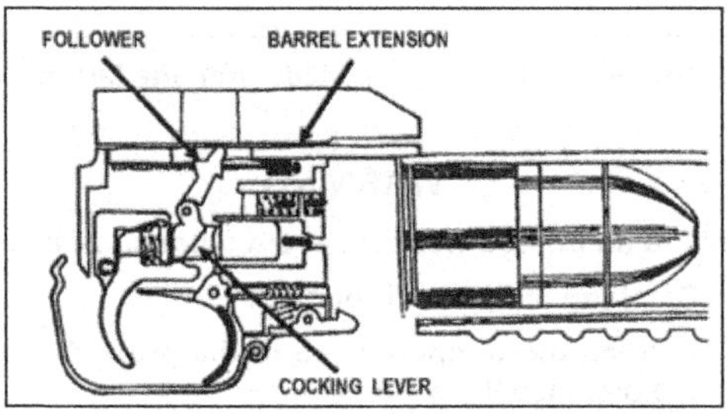

Figure 3-5 . Cocking the M203 grenade launcher.

c. **Extracting**. Extracting and cocking occur at the same time. As the grenadier opens the barrel assembly, a spring-loaded extractor keeps the live round or spent cartridge case seated against the receiver until the barrel clears the cartridge case (Figure 3-6).

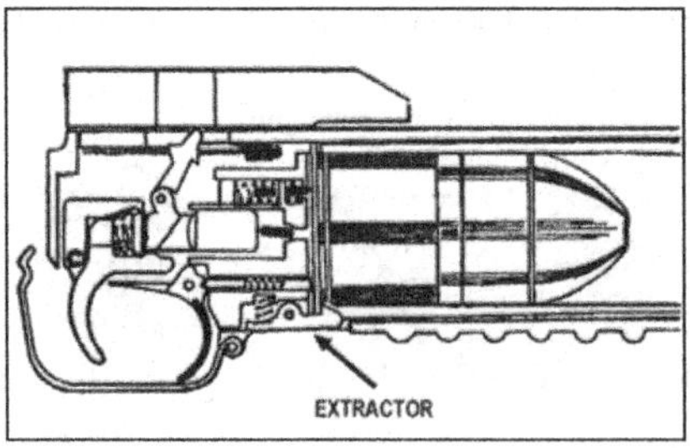

Figure 3-6 . Extracting the round or cartridge case.

d. **Ejecting**. The spring-loaded ejector pushes the live round or spent cartridge case from the barrel assembly (Figure 3-7).

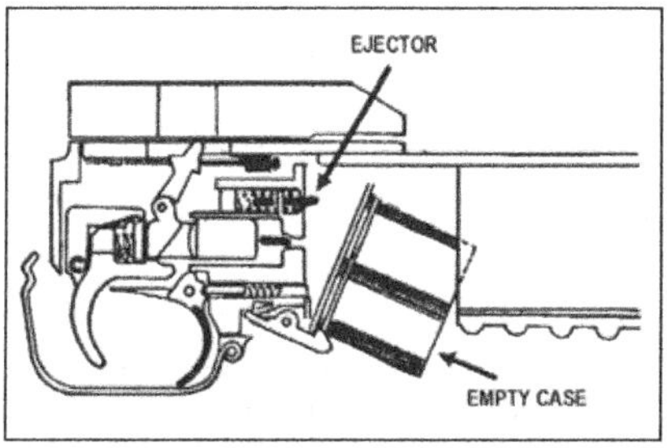

Figure 3-7 . Ejecting the round or cartridge case.

e. **Loading**. With the barrel assembly open, the grenadier inserts a round into the breech end of the barrel (Figure 3-8).

Figure 3-8 . Loading the M203 grenade launcher.

f. **Chambering**. As the grenadier closes the breech end of the barrel assembly, the extractor contacts the rim of the cartridge and seats (chambers) the round firmly (Figure 3-9).

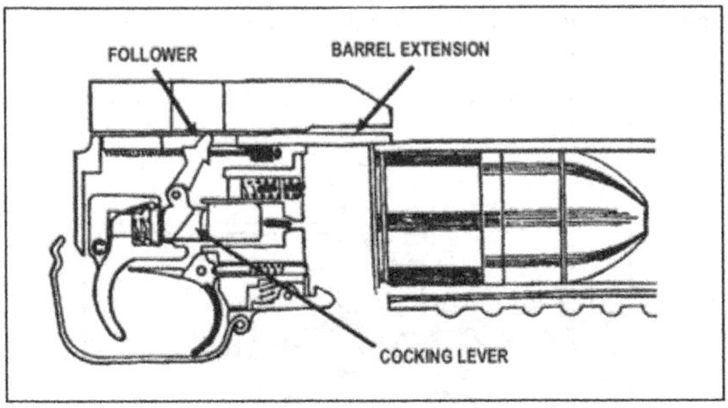

Figure 3-9 . Chambering a round.

g. **Locking**. As the barrel assembly closes, the barrel latch engages it. The cocking lever engages the barrel extension so that it cannot move forward along the receiver assembly.

h. **Firing**. When the grenadier pulls the trigger, the primary trigger sear disengages from the bottom sear surface of the firing pin. This releases the spring-driven firing pin, forcing it forward against the cartridge primer (Figure 3-10).

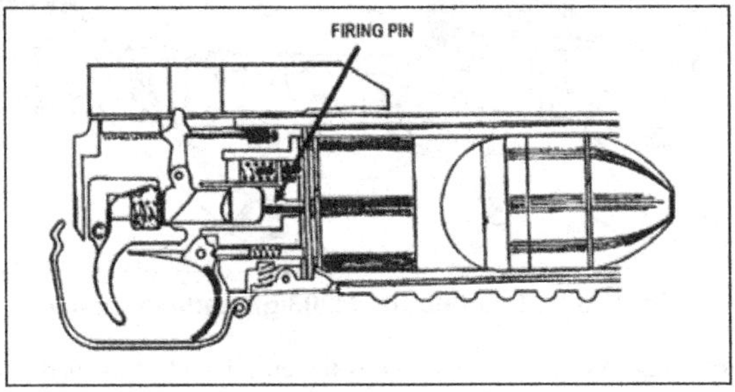

Figure 3-10. Firing the M203 grenade launcher.

This page intentionally left blank.

CHAPTER 4

PERFORMANCE PROBLEMS AND DESTRUCTION

This chapter identifies some of the problems that can cause the M203 grenade launcher to perform incorrectly. It also explains how to identify unserviceable parts and how to destroy the weapon when authorized to do so.

4-1. MALFUNCTIONS

A malfunction occurs when a mechanical failure prevents the weapon from firing properly. Neither defective ammunition nor improper operation of the weapon by the firer is a malfunction. The weapon should be cleaned, lubricated, and retried. If it still fails to function, it should be turned in to the unit armorer. Table 4-1 shows probable causes and corrective action for each type of malfunction.

Malfunction	Probable Cause	Corrective Action
Failure to cock	Broken sear	Notify unit maintenance
	Improper assembly of cocking lever	
	Loose, broken, or missing cocking lever spring pin	
Failure to lock	Excess plastic on breech end of barrel assembly	

Table 4-1. Malfunctions.

4-2. STOPPAGES

A stoppage is an unintentional interruption in the cycle of operation or functioning that may be cleared by immediate action. A stoppage is classified by its relationship to the cycle of functioning. Table 4-2 on page 4-2 shows the types of stoppages.

Stoppage	Probable Cause	Corrective Action
Failure to fire	Safety on	Place in fire position
	Empty chamber	Load
	Faulty ammunition	Reload
	Water or excess lubricant in firing pin well	Hand cycle weapon several times, to include pulling the trigger
	Worn or broken firing pin	Notify unit maintenance
	Dirt or residue in firing pin recess	Clean
	Blurred sear or firing pin	Notify unit maintenance
	Dirty firing pin well opening	
	Weak or broken firing pin spring	
Failure to extract	Defective extractor on spring or spring pin	
	Ruptured cartridge case	Remove from barrel
Failure to eject	Worn, broken, or missing ejector spring or retainer	Notify unit maintenance
Failure to chamber	Faulty ammunition	Reload
	Dirty chamber	Clean bore and chamber
Safety fails to stay in position	Missing spring pin or broken or worn safety	Notify unit maintenance

Table 4-2. Stoppages.

*4-3. IMMEDIATE ACTION

Immediate action refers to anything a Soldier does to reduce a stoppage without taking time to look for the cause. Immediate action should be taken in the event of either a hangfire or misfire. Either can be caused by an ammunition defect or by a faulty firing mechanism. Any failure to fire must be considered a hangfire until that possibility is eliminated.

a. What is the difference between a hangfire, a misfire, and a stoppage?

(1) A *hangfire* is a delay in the functioning of the round's propelling charge explosive train at the time of firing. The length of this delay is unpredictable, but in most cases, it ranges between a split second and 30 seconds. Such a delay in the functioning of the round could result from the presence of excess oil or grease, grit, sand, frost, or ice.

(2) A *misfire* is a complete failure of the weapon to fire. A misfire in itself is not dangerous, but because it cannot be immediately distinguished from a hangfire, it must be considered to be a hangfire until proven otherwise.

(3) A *stoppage* may be caused by a hangfire, the following precautions must be observed until the round has been removed from the weapon and the cause of the failure determined:

(a) Keep the M203 pointed downrange or at the target and keep everyone clear of its muzzle. If the stoppage occurs during training, shout MISFIRE and clear the area of any Soldiers not needed for the operation.

(b) Wait 30 seconds from the time of the failure before opening the barrel assembly. Once the 30 seconds has elapsed, open the barrel assembly slowly to about one-half inch or until you hear a metallic click. As soon as you hear the click, place the weapon on safe. You are now ready to perform the unloading procedure.

(c) After removing the round from the receiver, determine whether the round or the firing mechanism is defective. Examine the primer to see if it is dented. If the primer is dented, separate the round from other ammunition until it can be disposed of properly. However, if the primer is not dented, the firing mechanism is at fault. Once the cause of the failure to fire has been corrected, the round may be reloaded and fired.

WARNING

If you are unloading a weapon that has not been fired, avoid detonation either by catching the ejected round or by holding the weapon close to the ground to reduce the distance the round can fall.

4-4. REMEDIAL ACTION

Remedial action is any action taken by the gunner to restore his weapon to operational condition. Take remedial action only if immediate action does not remedy the problem.

4-5. DESTRUCTION PROCEDURES

Destruction of any military weapon is authorized only as a last resort to prevent the enemy from capturing or using it. This paragraph discusses planning for destruction, priorities and methods of destruction, and degree of damage. In combat situations, the commander has the authority to destroy weapons, but he must report doing so through channels.

a. **Planning**. SOPs for all units should contain a plan for destroying equipment. Having such a plan ensures that the damage is effective enough to deny use of the equipment to the enemy. The plan must be flexible enough in its designation of time, equipment, and personnel to meet any situation.

b. **Priorities of Destruction**. When lack of time prevents them from completely destroying equipment, Soldiers must destroy the same essential parts on all like equipment. The order in which the parts should be destroyed (priority of destruction) follows:

(1) Bolt assembly (M16) and breech mechanism (M203).

(2) Barrels (both M16 and M203).

(3) Sights or sighting equipment (including nightsight).

(4) Optics mount.

c. **Methods of Destruction**. Equipment may be destroyed by any of several methods. The commander must use his imagination and resourcefulness to select the best method of destruction based on the facilities available. Time is usually critical. The methods of destruction are as follows:

(1) ***Mechanical***. Use an axe, pick, sledgehammer, crowbar, or other heavy implement.

(2) ***Burning***. Use gasoline, oil, incendiary grenades, other flammables, or a welding or cutting torch.

(3) ***Demolition***. Use suitable explosives or ammunition or, as a last resort, hand grenades.

(4) ***Disposal***. Bury essential parts, dump them in streams, or scatter them so widely that recovering them would be impossible.

d. **Degree of Damage**. The method of destruction used must damage equipment and essential spare parts to the extent that they cannot be restored to usable condition in the combat zone, either by repair or by cannibalization.

CHAPTER 5

MARKSMANSHIP TRAINING

**Marksmanship training is conducted in three phases. Every phase has the same three objectives: to teach each grenadier--*

- *To hit the target accurately with the first round,*
- *To adjust fire, and*
- *To do both quickly.*

This chapter discusses the first two phases: ***preliminary marksmanship training****, which develops nonfiring individual skill proficiency (Section I), and* ***basic gunnery*** *(Section II). In these two phases, the Soldier learns to apply the fundamentals of gunnery and to zero the M203 during qualification exercises in day, CBRN, and night conditions. (Chapter 6 discusses the third phase,* ***advanced gunnery****.)*

***WARNING**

Before allowing anyone to move between stations, ensure that--

- **All rifles and grenade launchers have been cleared.**
- **Bolts are to the rear.**
- **Barrel assemblies are in the open position.**

Anyone observing an unsafe act should call out "Cease fire" and notify range personnel immediately.

Section I. PRELIMINARY MARKSMANSHIP TRAINING

Grenadiers and leaders must master marksmanship fundamentals before firing individually or collectively. During preliminary marksmanship training, grenadiers learn and demonstrate the individual skills that prepare them to fire live ammunition. After learning the characteristics and mechanics of the weapon (Chapters 2, 3, and 4), they learn the four fundamentals of marksmanship, sight manipulation, and response to fire commands. Dry-fire exercises are excellent for training to proficiency. Good preliminary marksmanship instruction improves individual proficiency, which in turn improves the proficiency of collective fire.

*5-1. FOUR FUNDAMENTALS OF MARKSMANSHIP

The four fundamentals of M203 marksmanship are steady position, aiming, breathing, and trigger control. Only the first fundamental (steady position) varies. The other three remain the same regardless of the Soldier's position.

a. **Steady Position**. This varies according to the position and the type of sight used (quadrant or leaf). Note that for firing prone, a supported position is best.

(1) ***Prone Position, Quadrant Sight*** (top, Figure 5-1).

(a) Lie face down, grasp the M16 pistol magazine with your right hand, and place the butt of the rifle into the pocket of your right shoulder.

(b) Lower your right elbow to the ground so your shoulders are level. This places the weight of your body behind the weapon, which enables you to recover quickly each time you fire.

(c) Grasp the barrel grip with your left hand, supporting with sandbags. Straighten your upper body and spread your legs a comfortable distance apart. Try to point your toes outward and relax your ankles so your heels will rest on the ground. Relax the weight of your upper body forward onto your left arm.

(2) ***Prone Position, Leaf Sight*** (bottom, Figure 5-1). While firing with the leaf sight at ranges greater than 150 meters, place the butt stock of the weapon under your armpit and grip firmly to prevent the weapon from moving. Lean your head 45 degrees to the right and place the M16 front sight post on the desired range. Raise the butt stock and lower the muzzle to obtain the proper sight alignment and sight picture.

Figure 5-1. Prone supported position, quadrant sight (top) and leaf sight (bottom).

***WARNING**

Before allowing anyone to move between stations, ensure that the sling is clear of the weapon muzzle before firing.

(3) ***Kneeling Position*, *Quadrant Sight*** (left, Figure 5-2).

(a) Kneel on your right knee while facing the target with your right hand on the magazine and your left hand grasping the barrel grip.

(b) Place your left foot about .45 meter (18 inches) to your left front with your toes pointing in the general direction of the target.

(c) Keeping your right toe in place, sit on your right heel.

(d) Place your left elbow forward of your left knee, resting the flat portion of your upper arm on your knee.

(e) Move the rifle butt into the pocket of your right shoulder, pulling the rifle magazine with your right hand and grasping the barrel grip with your left hand.

(f) With your right hand on the rifle magazine, place your right forefinger in the trigger guard of the grenade launcher.

(g) Pull the rifle firmly into your shoulder.

(h) Pull your right elbow in close to your body to help you apply rearward pressure to the weapon. Ensure that your leg completes a solid, three-point base for your position.

(4) ***Kneeling Position*, *Leaf Sight*** (right, Figure 5-2).

(a) For ranges greater than 150 meters, place the butt stock of the weapon under your armpit and grip firmly to prevent the weapon from moving.

(b) Lean your head 45 degrees to the right and place the front sight post of the M16 on the desired range. Raise the butt stock and lower the muzzle to obtain the proper sight alignment and sight picture.

Figure 5-2. Kneeling position, quadrant sight (L) and leaf sight (R).

(5) ***Sitting Position, Open-Legged, Quadrant Sight*** (left, Figure 5-3, page 5-4).

(a) Sit down, breaking your fall with your right hand, and slide your buttocks well to the rear. Face the target half right, and spread your feet wide.

(b) Grasp the rifle magazine with your right hand and the barrel grip with your left hand.

(c) Bend forward from your hips.

(d) Move the butt of the rifle into the pocket of your right shoulder, still holding the rifle magazine with your right hand.

(e) Pull the weapon down slightly with your left hand and pull it to the rear firmly with your right hand.

(6) ***Sitting Position, Open-Legged, Lea f Sight*** (right, Figure 5-3).

(a) For ranges greater than 150 meters, place the butt stock of the weapon under your armpit and grip firmly to prevent the weapon from moving.

(b) Lean your head 45 degrees to the right and place the M16 front sight post on the desired range. Raise the butt and lower the muzzle to obtain the proper sight alignment and sight picture.

Figure 5-3. Sitting position, open-legged, quadrant sight (L) and leaf sight (R).

(7) ***Sitting Position, Cross-Ankle, Quadrant Sight*** (left, Figure 5-4).

(a) Sit facing the target half right.

(b) Extend your legs from your body and cross your left ankle over your right ankle.

(c) Keep both ankles straight.

(d) Grasp the rifle magazine with your right hand and the barrel grip with your left.

(e) Place your left upper arm across your left knee.

(f) Move the butt of the rifle into the pocket of your right shoulder.

(8) ***Sitting Position, Cross-Ankle, Leaf Sight*** (right, Figure 5-4).

(a) For ranges greater than 150 meters, place the butt stock of the weapon under your armpit and grip firmly to prevent the weapon from moving.

(a) Lean your head 45 degrees to the right and place the M16 front sight post on the desired range. Raise the butt and lower the muzzle to obtain the proper sight alignment and sight picture.

Figure 5-4. Sitting position, cross-ankled, quadrant sight (L) and leaf sight (R).

(9) ***Sitting Position, Cross-Legged, Quadrant Sight*** (left, Figure 5-5).

(a) Sit down facing the target half right.

(b) Cross your left leg over your right leg and draw both feet close to your body.

(c) Grasp the rifle magazine with your right hand.

(d) Move the butt of the rifle into the pocket of your right shoulder, and grasp the rifle barrel grip properly with your left hand.

(10) ***Sitting Position, Cross-Legged, Leaf Sight*** (right, Figure 5-5).

(a) For ranges greater than 150 meters, place the butt stock of the weapon under your armpit and grip firmly to prevent the weapon from moving.

(b) Lean your head 45 degrees to the right and place the M16 front sight post on the desired range. Raise the butt and lower the muzzle to obtain the proper sight alignment and sight picture.

Figure 5-5. Sitting position, cross-legged, quadrant sight (L) and leaf sight (R).

(11) ***Squatting Position, Cross-Legged, Quadrant Sight*** (left, Figure 5-6, page 5-6).

(a) Turn half right to the target and, keeping both feet flat on the ground and a comfortable distance apart, squat as low as you can.

(b) Grasp the rifle magazine with your right hand.

(c) Place your left upper arm inside your left knee and the butt of the rifle into the pocket of your right shoulder. Grasp the rifle barrel grip properly.

(d) Lower your right elbow against the inside of your right knee.

(12) ***Squatting Position, Cross-Legged, Leaf Sight*** (right, Figure 5-6, page 5-6).

(a) For ranges greater than 150 meters, place the butt stock of the weapon under your armpit and grip firmly to prevent the weapon from moving.

(b) Lean your head 45 degrees to the right and place the M16 front sight post on the desired range. Raise the butt and lower the muzzle to obtain proper sight alignment and sight picture.

Figure 5-6. Squatting position, quadrant sight (L) and leaf sight (R).

(13) ***Fighting Position, Quadrant Sight*** (left, Figure 5-7).

(a) If possible, use support when firing from a fighting position.

(b) Place your right foot against the rear of the fighting position and lean forward until your chest is against its forward edge.

(c) Grasp the magazine with your right hand.

(d) Place your left elbow on or against solid support.

(e) Use your right hand to position the butt of the rifle in the pocket of your right shoulder. Grasp the rifle barrel grip properly.

(f) Place your right elbow on or against a solid support and relax into a comfortable firing position.

(14) ***Fighting Position, Leaf Sight*** (right, Figure 5-7).

(a) For ranges greater than 150 meters, place the butt stock of the weapon under your armpit and grip firmly to prevent the weapon from moving.

(b) Lean your head 45 degrees to the right and place the M16 front sight post on the desired range. Raise the butt and lower the muzzle to obtain proper sight alignment and sight picture.

Figure 5-7. Fighting position, quadrant sight (L) and leaf sight (R).

NOTE: The weapon must not touch the support.

(15) ***Standing Position, Quadrant Sight*** (left, Figure 5-8).

(a) Face the target while standing with your feet spread a comfortable distance apart.

(b) Grasp the rifle barrel grip with your left hand and the rifle magazine with your right hand.

(c) Place the butt of the stock into your right shoulder so that the sight is level with your eyes.

(d) Hold your right elbow high to form a good pocket for the butt of the stock and to permit a strong rearward pressure with your right hand.

(e) Hold most of the weight of the weapon with your left hand.

(f) Shift your feet until you achieve a natural aiming stance.

(16) ***Standing Position, Leaf Sight*** (right, Figure 5-8).

(a) For ranges greater than 150 meters, place the butt stock of the weapon under your armpit and grip firmly to prevent the weapon from moving.

(b) Lean your head 45 degrees to the right and place the M16 front sight post on the desired range. Raise the butt and lower the muzzle to obtain the proper sight alignment and sight picture.

Figure 5-8. Standing position, quadrant sight (L) and leaf sight (R).

b. **Aiming**. Aiming procedures for every position follow:

(1) ***Aligning Sight***. When using the leaf sight, align it with the front sight post of the M16. When using the quadrant sight, align its rear sight aperture with its front sight post. Picture a horizontal line through the center of the leaf sight or rear sight aperture: the top of the M16's front sight post should touch this line. Picture a vertical line through the center of the leaf sight or rear sight aperture: this line should vertically bisect the front sight post (Figure 5-9, page 5-8).

(2) ***Focusing***. For either sight, focus on the front sight post. A good firing position places your eye directly on line with the center of the leaf sight or rear sight aperture. Your eye's natural ability to center objects in a circle and seek the point of greatest light will help you align the sight correctly.

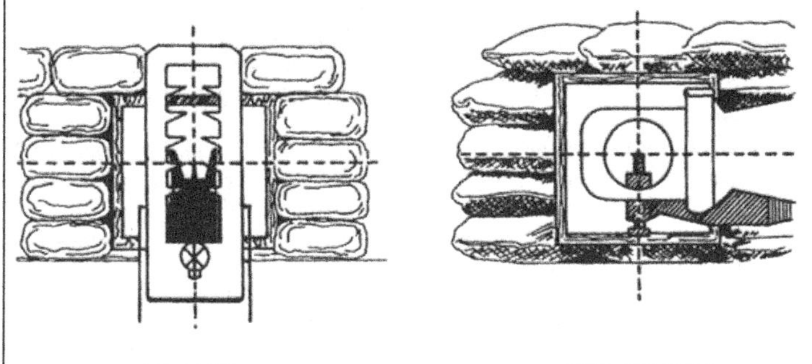

Figure 5-9. Sight pictures for leaf and quadrant sights.

(3) ***Obtaining Sight Picture***. To achieve a correct sight picture, align the front sight post and the leaf sight or rear sight aperture with the target. For area targets, aim where the round's bursting radius will make the round most effective. For point targets, aim at the target's center of mass.

c. **Breathing**. The technique for breathing is the same for every position: Breathe naturally, exhale most of your air, hold your breath, and fire before you become uncomfortable. In combat, just choke off your breath before firing.

d. **Trigger Control**. The technique for trigger control is the same for every position. Place your trigger finger (the index finger of your right hand) so that the trigger is between the first joint and the tip of your finger (not at the extreme end of your finger). Adjust for your hand size and grip. Then, squeeze your trigger finger to the rear without disturbing the lay of the weapon.

5-2. LIMITED VISIBILITY

The fundamentals of marksmanship are almost the same in limited visibility as in normal visibility.

a. **Steady Position**. An M203 with an AN/PVS-4 mounted on it leans to the left. When assuming a steady position, the grenadier must apply more rearward pressure to compensate for the lean and then steady the weapon.

b. **Aiming**. The grenadier sights with the reticle of the AN/PVS-4 rather than with the M203's iron sights. Sighting this way requires him to change position, which breaks his stock weld and makes the weapon seem heavier.

c. **Breathing**. Though breathing itself is affected little by limited visibility, using night vision devices that magnify the field of view increases the effect of weapon movement caused by breathing.

d. **Trigger Control**. This is the same regardless of visibility conditions. The objective is to keep the weapon aligned with the target.

*e. **Night Vision Devices**. The AN/PVS-7 is issued for use with the M203, whereas the AN/PVS-4 is normally issued for use with crew-served weapons. M203 gunners may qualify with either device. In a defensive position, the gunner identifies targets during daylight and constructs aiming or elevation stakes. Because the AN/PVS-7 rear sight must be set to the far setting to sense rounds, the gunner cannot see both the M203 sights and the target at the same time. Therefore, stakes are more important with the AN/PVS-7 than with the AN/PVS-4.

f. **Marked-Sling Method**. The best field-expedient method for firing the M203 grenade launcher in limited visibility is the marked-sling method using only the M16 rifle series (Figure 5-10, page 5-10). To use this method, the grenadier must--

(1) Face the target and kneel on the right knee (if firing right-handed), keeping the left foot pointed toward the target.

(2) Loosen the sling and place the forward foot in the sling.

(3) Place the butt of the stock firmly on the ground.

(4) Using the left hand, grasp firmly the upper barrel grip just below the barrel.

(5) Grasp the receiver group with the right hand.

WARNING

Placing your knee against the buttstock could injure your knee.

(6) Ensure the sling is taut and vertical between the front sling swivel and the boot (Figure 5-11, page 5-10). If not, the rounds will impact at a greater range than desired.

(7) Fire several rounds to determine the desired range.

(8) Mark the sling (where it is held to the ground by the foot) with colored tape, paint, ink, or whatever is available. Mark the position of the buckles (Figure 5-12, page 5-10) so that, if either is moved, the grenadier can return them to their original positions and be assured of constant range accuracy.

(9) If the sling gets wet, it may stretch or shrink, indirectly causing the rounds to impact closer or farther than desired.

*5-3. CHEMICAL, BIOLOGICAL, RADIOLOGICAL, AND NUCLEAR ENVIRONMENTS

The fundamentals of marksmanship remain valid in the CBRN environment, but some modifications may be needed to accommodate the equipment.

a. **Steady Position**. Bulky CBRN wear requires the grenadier to press the stock of the weapon more firmly into his shoulder pocket.

b. **Aiming**. Aiming is affected little by CBRN.

c. **Breathing**. Wearing the protective mask makes breathing more difficult. Grenadiers must try to breathe normally to avoid hyperventilating while firing.

d. **Trigger Control**. All Soldiers must wear rubber gloves.

Figure 5-10. Marked-sling method.

Figure 5-11. Front-sling swivel and front of boot.

Figure 5-12. Marked sling and buckle highlighted.

5-4. FIRE COMMANDS

Standard fire commands are explained to grenadiers and are used during all subsequent gunnery training. Trainers give the appropriate elements before each dry-fire or live-fire exercise. The grenadier performs as directed and repeats each element as it is announced. (Chapter 6 discusses fire commands in detail.)

a. **Alert**. The trainer gives the alert as a fire mission. On hearing this, the grenadier loads the weapon and moves the safety lever to "F."

b. **Direction**. The trainer gives the direction to target.

c. **Description**. The trainer describes the target, for example, "Bunker" or "Machine gun position," and the grenadier lays on the target.

d. **Range**. The trainer gives the (estimated) range to the target, for example, "150."

e. **Method of Fire**. The method of fire for either target is three rounds. On the basic range, grenadiers fire at both point and area targets.

f. **Command to Open Fire**. To open fire, the trainer commands COMMENCE FIRING or AT MY COMMAND. When ready, the grenadier announces "Up" and fires or waits for the command to fire. When all grenadiers are ready, the trainer gives the actual command to fire.

5-5. DRY-FIRE EXERCISES

Dry-fire exercises train grenadiers in the techniques of loading, unloading, immediate action, fundamentals of marksmanship, and sight manipulation. These exercises are conducted with TP or dummy rounds. The trainer gives fire commands as appropriate.

a. **Loading and Unloading Exercise**. This trains the grenadier to operate and clear the weapon proficiently. Loading and unloading procedures (Chapter 3) should be practiced with dummy ammunition.

b. **Immediate Action Exercise**. This exercise is conducted with a dummy round and the basic grenade launcher target.

(1) Load the weapon with a dummy round and aim it at one of the targets on the basic grenade launcher range.

(2) Maintain the sight picture while you pull the trigger to simulate firing.

(3) When you are informed that you have a misfire, apply misfire procedures; then continue to fire (Chapter 4).

c. **Aiming Exercise**. This exercise requires the grenadier to simulate firing a dummy round at a target on the basic grenade launcher range.

(1) Maintain your sight picture throughout the firing cycle.

(2) If, after firing, you note that the sight picture has moved, then you were unsteady when you fired.

(3) After each shot, apply immediate action procedures to extract and eject the dummy cartridges. Then recock the barrel assembly.

d. **Sight Setting and Sight Changing Exercises**. These exercises train the grenadier to operate and adjust both quadrant and leaf sights.

(1) ***Range***. Manipulate the sights to different range settings (quadrant sight, 50 to 400 meters; leaf sight, 50 to 250 meters). To learn to make fine adjustments for elevation, manipulate the sights from the minimum to the maximum setting. When you do not have time to adjust the sights, you may adjust the aiming point instead.

(2) ***Windage***. Depress the rear sight aperture left and right and traverse the windage screw across the entire scale.

e. **Dry-Fire Proficiency (Performance) Exam**. Grenadiers practice the dry-fire tasks until they become proficient in operating the weapon; then they take the dry-fire proficiency exam (Appendix C). This exam emphasizes learning by doing. Before he can progress to live firing, each grenadier must demonstrate skill in every task in the exam.

f. **Remedial Training**. Soldiers who fail to pass the performance exam must attend remedial training, after which they retest. Those who pass can help train those having difficulty.

5-6. SENSING AND ADJUSTMENT OF FIRE

The grenadier determines (senses) where the grenade landed relative to the target and then adjusts elevation and deflection.

a. **Sensing**. As soon as the grenade explodes, determine where it exploded with respect to the target. This is called "sensing" (the impact) and has two aspects: range and deviation. Because the casualty radius of the HE round is 5 meters (5 1/2 yards), determine both range and deviation to the nearest 5 meters.

(1) ***Range***. Sense the range as one of the following:

(a) *Short*. The grenade bursts between you and the target.

(b) *Over*. The grenade bursts beyond the target.

(c) *Target*. The grenade hits any part of the target.

(d) *Range Correct*. The grenade bursts slightly left or right of the target, but at the correct range.

(e) *Doubtful*. The grenade burst left or right of the grenadier, but you cannot sense the range.

(2) ***Deviation***. Announce a deviation sensing as either--

- Right or left of the target, or
- On line with the target.

b. **Adjustment of Fire**. To ensure a second-round hit, adjust your fire by sensing the impact of the round and manipulating the sight.

(1) If time allows, whether using the AN/PVS-4 or AN/PVS-7, adjust the sights; if time is critical, adjust the point of aim instead.

(2) If the grenade lands more than 25 meters over or short of the target, adjust the range quadrant to bring the next grenade on target.

(3) If the grenade explodes less than 25 meters from the target, adjust the point of aim to bring the next grenade on target.

(4) If the launcher is properly zeroed, deviation errors are normally small and easily corrected by adjusting the aiming point. A wind strong enough to move the grenade out of its normal trajectory, however, increases the size of the deviation errors. After observing the effect of the wind on the strike of the grenade, compensate for the effect of the wind by aiming into it. This should help bring the next grenade on target. For example, if the grenade bursts to the left and short of the target, sense the strike of the round relative to the target and then adjust an equivalent distance to the right and over the target to achieve a target hit. Watch the flight of the grenade to the target. This helps determine the effect of the wind on the grenade as it moves toward the target. Evaluating and compensating for the wind before firing increases your chances of achieving a first-round hit.

*5-7. LAYOUT OF GRENADE LAUNCHER RANGE

The grenade launcher range is designed for all grenade launchers. Because Soldiers can qualify on this range in all conditions, it prepares grenadiers for combat situations. The range has four stations (Figure 5-13). Minimum range personnel and their duties are the same for M203 qualification firing as they are for other grenadier firing. These personnel include an OIC, NCOIC, safety officer, ammunition NCO, tower operator, station NCOs, primary trainer, and concurrent training trainers. However, local policy may require more personnel. (Appendix D discusses range safety; TC 25-8 provides a detailed setup and target configuration for this range.)

5-8. DESCRIPTION OF RANGE AND TARGETS

The range has four self-contained stations (Figure 5-13). It is 30 meters wide by 500 meters deep and has a no-HE fire zone out to 165 meters. (Grenadiers can fire HE only on Stations 1, 3, and 4.) Targets should be built from durable materials to reduce downrange target maintenance. Those within each station must be grouped and spaced so that the grenadier may fire on close-range, mid-range, and long-range targets, in that order. The following description of the stations and targets is included to help trainers maintain control during zeroing, practice, and record fire. Figure 5-14 shows the range in meters and feet for combat, point target, training, area target and maximum ranges for the M203.

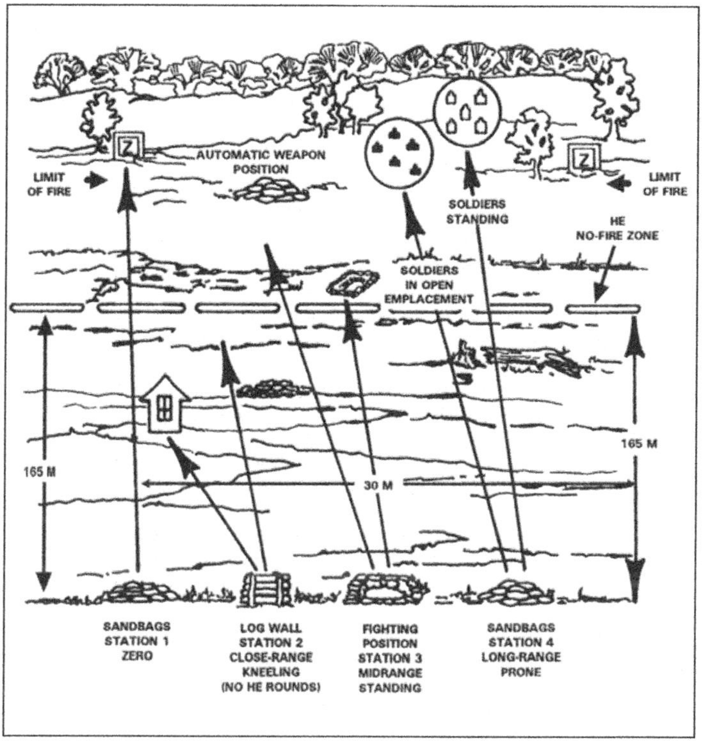

Figure 5-13. Grenade launcher range.

WARNING

Before firing HE grenades, move to a protected (covered) position. The minimum safe ranges follow:

Training	165 meters	(541 feet)
Combat	31 meters	(102 feet)

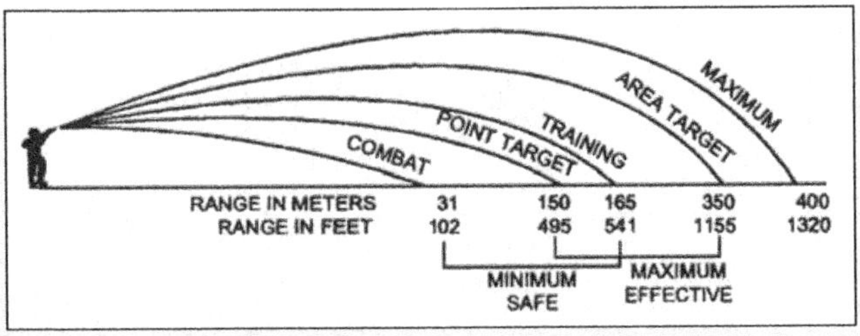

Figure 5-14. Ranges in meters and feet.

NOTE: To simplify the task of finding and destroying duds, trainers must ensure the impact area is free of any vegetation other than short grass.

a. Station 1 consists of a prone fighting position with a log or sandbag support and a zeroing target at 200 meters. The target should be constructed of logs or other suitable material. It must have a surface at least 2 meters high by 2 meters wide (6 feet by 6 feet). The target should be clearly marked with a large "Z" painted in a color that contrasts with the surrounding background and that is visible in different sun or glare conditions.

b. Station 2 consists of an upright log or log wall, a kneeling firing position about 4 feet high, and 2 point-type targets. The targets include a simulated window or door of a building at 90 to 100 meters and a small bunker or fighting position with overhead cover at 125 meters. The targets may be constructed of logs, sandbags, or other suitable material.

c. Station 3 consists of a fighting position and two targets. The targets are a two-person bunker at 150 meters and an automatic weapon position at 175 meters. The bunker represents a point target, and the automatic weapon position represents a target that can be engaged with area-type fire. The targets may be constructed of logs, sandbags, or other suitable material.

d. Station 4 consists of a prone fighting position with a log or sandbag support and two area-type targets (with personnel targets in the open) at 250 and 350 meters. The log or sandbags at the firing position are used for support and cover. The targets are E-type and F-type shaped silhouette and are made out of durable materials.

Section II. BASIC GUNNERY

Basic gunnery allows the grenadier to zero and apply the fundamentals of marksmanship during live-fire exercises in day, night, and CBRN conditions.

5-9. ZEROING OF M203 GRENADE LAUNCHER

A correct zero consists of the elevation and windage sight settings that enable the grenadier to hit the point of aim at a given range with one of the three sighting systems: leaf, quadrant, or night sight. To zero the M203 using either the leaf sight or quadrant sight, the grenadier engages a target at 200 meters. (The M203 is normally zeroed using only the quadrant sight, but may be zeroed with both sights or with only the leaf sight.)

a. **Zeroing the Leaf Sight**. A red mark at 50 meters on the leaf sight reminds the grenadier not to zero at this range.

(1) Select a target at 200 meters.

(2) Place the sight in the upright position.

(3) Place the center mark of the windage scale on the index line on the rear of the sight base.

(4) Loosen the elevation adjustment screw on the leaf sight.

(5) Place the leaf sight's index line on the sight mount's center elevation mark.

(6) Tighten the elevation adjustment screw.

(7) Assume a prone supported firing position.

(8) Load one round of 40-mm HE or TP ammunition.

(9) Use correct sighting and aiming procedures to align the target with the front leaf sight.

(10) Fire a round, sense the impact, and adjust the sight.

(a) *Windage*. Turn the sight windage screw clockwise to move the leaf sight to the left and vice versa. One increment moves round impact 1 1/2 meters at a range of 200 meters.

(b) *Range*. Use a 40-mm cartridge case and turn the elevation adjustment screw to raise the leaf sight (this increases range) or to lower the leaf sight (this decreases range). Moving the leaf sight one increment moves the impact of the round 10 meters at a range of 200 meters.

(11) Fire two more cartridges, readjusting the sight after each. Once a round impacts within 5 meters of the target, the weapon is zeroed.

(12) After you have zeroed the weapon, record the zero data on your scorecard. As soon as you can, transfer the information to a small piece of paper and tape this inside the M16 pistol grip.

b. **Zeroing the Quadrant Sight**.

(1) Select a target at 200 meters.

(2) Ensure that the quadrant sight is correctly mounted on the rifle's carrying handle.

(3) Open the front sight post and rear sight aperture.

(a) Move the front sight post to its highest position and then back 2 1/2 turns.

(b) Depress the rear sight retainer. Slide the rear sight aperture to the left until its white index line aligns with the edge of the sight aperture arm.

(4) Move the sight latch rearward and reposition the quadrant sight arm to zeroing range (200 meters).

(5) Assume a prone supported firing position.

(6) Use correct sighting and aiming procedures to align the target with the front sight post and rear sight aperture.

(7) Load one round of 40-mm HE or TP ammunition.

(8) Fire a round, sense the impact, and adjust the sight.

(a) *Elevation*. Turn the front sight post right (to decrease elevation) or left (to increase elevation). At a range of 200 meters, one full turn equals 5 meters.

(b) *Windage*. Press the sight aperture retainer. Move the rear sight aperture away from the barrel to move the trajectory to the left or vice versa. At a range of 200 meters, one notch on the rear sight aperture equals 1 1/2 meters.

(9) Fire two more cartridges, readjusting the sights after each. If the round lands within 5 meters of the target, the weapon is zeroed.

(10) After you have zeroed the weapon, record the zero data on your scorecard. As soon as you can, transfer this information to a small piece of paper and tape this inside the M16 pistol grip.

5-10. OVERALL QUALIFICATION STANDARDS

DA Form 2946-R (40-mm Grenade Launcher Scorecard) is used for qualification firing. To qualify with an M203, a grenadier must perform to prescribed standards and must score at least 60 of 90 possible points. Each target hit is worth 10 points. Zeroing is not included on the scorecard because the weapon must be zeroed before qualification firing. However, the zero data should already have been entered on the scorecard when the weapon was zeroed. HE familiarization may be included in qualification firing but is not scored. Ratings are awarded based on the point chart shown on the scorecard. A blank scorecard is provided in back of this field manual and must be locally reproduced on 8 1/2- by 11-inch paper. Figure 5-15, page 5-18 shows an example of a completed scorecard.

*5-11. DAY RECORD FIRE

Day record fire gives the grenadier the confidence and experience he needs to progress from dry-fire exercises to record fire. All Soldiers must be prepared to accomplish their missions, even in protective clothing; thus, day record fire includes two CBRN tasks (Tasks 4 and 5). This exercise is conducted on a grenade launcher range IAW Firing Table I (Table 5-1, page 5-20). Before they fire for qualification, grenadiers must first zero their weapons and receive instruction on the objectives, range, targets, and qualification standards. The unit is organized in firing orders based on range constraints. Each firing order consists of two grenadiers, one of whom assists. Grenadiers fire this exercise from the following fighting positions: kneeling supported, mid-range supported, long-range supported, CBRN mid-range point target, and CBRN mid-range area target. For each of these tasks, the grenadier can designate which target he will engage first.

40-MM GRENADE LAUNCHER SCORECARD

For use of this form, see FM 3-22.31; the proponent agency is TRADOC.

1. ID CODE *	2. UNIT	3. DATE *(YYYYMMDD)*
20070419BTD01	C CO, 2 29 IN	20070419

4. ZERO LEAF SIGHT	DEFILADE	ELEVATION	5. ZERO QUADRANT SIGHT	DEFILADE	ELEVATION
				L4	R1

6. RECORD FIRE	TASK NO.	TIME (MIN)	TGT	HIT	MISS
DAY	1	2	1	X	
			2		X
	2	2	1	X	
			2	X	
	3	2	1	X	
			2	X	
CBRN	4	2	1	X	
	5	2	1	X	
NIGHT	6	2	1	X	
* Do not use personal information	7. TOTAL POINTS *(Award 10 points for each hit)*			8	

8. RATING SCALE

TOTAL POINTS *(Block 7)*	RATING	
80 - 90	EXPERT	[X]
70 - 75	FIRST CLASS	[ ]
60 - 65	SECOND CLASS	[ ]
0 - 55	UNQUALIFIED	[ ]

9. GRADER'S INITIALS	10. DATE *(YYYYMMDD)*
RM	20070419
11. OIC'S INITIALS	**12. DATE *(YYYYMMDD)***
CS	20070419

EXAMPLE

DA FORM 2946-R, FEB 2007 EDITION OF NOV 2002 IS OBSOLETE. APD V1.00

Figure 5-15. Example completed DA Form 2946-R, 40-mm Grenade Launcher Scorecard.

WARNING

Before allowing anyone to move between stations, ensure that--

- **All weapons are clear.**
- **Bolts are to the rear.**
- **Barrel assemblies are open.**

If you see an unsafe act, call out "Cease fire" and notify range personnel immediately.

a. **Station 1, Zeroing**. The grenadier zeroes with both quadrant and leaf sights at Station 1.

(1) ***Leaf Sight***. From a prone supported firing position, fire to zero the weapon. This reinforces the experience gained during dry firing and allows practice in loading and firing with the most accurate sensing and adjustments obtainable. If you zero in three rounds, use the other two rounds to confirm the zero. If you cannot zero with five rounds, the trainer must remove you from the firing line for remedial training.

(a) Prepare the sight for zeroing.

(b) Assume a good prone supported firing position.

(c) When you receive the following fire command, repeat each element as it is given:

GRENADIER
FRONT
200 (ZERO PANEL)
ONE ROUND
COMMENCE FIRING

(d) Load one round, obtain the proper sight picture, and announce "Up" to your assistant.

(e) When the tower operator gives the command to commence firing, fire one round at the panel marked "Z."

(f) Sense the impact of the round. If the round did not land within 5 meters of the zero panel, adjust the sights for windage and elevation.

(g) Repeat until a round lands within 5 meters of the zero panel.

(h) Once you have zeroed the weapon, record the zero data on your scorecard. As soon as you can, transfer the information to a small piece of paper and tape this inside the M16 pistol grip.

(2) ***Quadrant Sight.*** From a prone supported firing position, fire to zero the weapon. This reinforces the experience gained during dry firing and gives you practice in loading and firing with the most accurate sensing and adjustments you can obtain. The steps for zeroing with the quadrant sight are the same as those for zeroing with the leaf sight.

b. **Station 2, Task 1, Kneeling Supported Position**. (Only TP rounds may be used at this station.)

(1) When you receive the command DESIGNATE THE TARGET, identify the target you intend to engage by announcing WINDOW or BUNKER.

(2) When you receive the command DETERMINE THE RANGE, announce the range to the target.

(3) Load one of the three rounds allotted. Because HE must not be fired at ranges of less than 165 meters on the basic grenade launcher range, use only TP rounds.

(4) When you receive the following fire command, repeat each element as it is given:

GRENADIER
FRONT
3 ROUNDS
90 meters (window) or 100 meters (bunker)
COMMENCE FIRING

(5) Acquire the proper sight picture and announce "Up" to the grader.

(6) Engage the target given in the fire command until you hit it. Fire any remaining rounds at the second target. You need no other fire command. For each round you fire, your assistant announces "Hit" or "Miss."

STATION	TASK	TIME	ROUNDS	TYPE	TARGET(S) AND RANGE(S)
2	1	2 min	3	TP	Window at 90 to 100 meters Bunker at 105 to 125 meters
3	2	2 min	3	HE	Bunker at 170 to 180 meters Automatic weapon at 190 to 200 meters
4	3	2 min	3	HE	Troops in open emplacement at 275 to 300 meters Troops in open at 235 to 350 meters
3	4 CBRN	2 min	3	HE	Bunker at 170 to 180 meters
3	5 CBRN	2 min	3	HE	Automatic weapons at 190 to 200 meters

Table 5-1. Firing Table I, day record fire qualification.

c. **Station 3, Task 2, Midrange Supported Position**.

(1) When you receive the command DESIGNATE THE TARGET, identify the target you intend to engage by announcing "Bunker" or "automatic weapon."

(2) When you receive the command DETERMINE THE RANGE, announce the range to the target.

(3) Load one of the three rounds allotted.

(4) When you receive the following fire command, repeat each element as it is given:

GRENADIER
FRONT
3 ROUNDS
170 METERS (BUNKER) OR 190 METERS (AUTOMATIC WEAPON)
COMMENCE FIRING

(5) Acquire the proper sight picture and announce "Up" to the grader.

(6) Engage the target given in the fire command until you hit it. Fire any remaining rounds at the second target. You need no other fire command. For each round you fire, your assistant announces "Hit" or "Miss."

d. **Station 4, Task 3, Long-Range Supported Position**.

(1) When you receive the command DESIGNATE THE TARGET, identify the target you intend to engage by announcing "Troops in the open emplacement" or "Troops in the open."

(2) When you receive the command DETERMINE THE RANGE, announce the range to the target.

(3) Load one of the three rounds allotted.

(4) When you receive the following fire command, repeat each element as it is given:

GRENADIER
FRONT

3 ROUNDS
300 METERS (TROOPS IN THE OPEN EMPLACEMENT) or
 350 METERS TROOPS IN OPEN
COMMENCE FIRING

(5) Acquire the proper sight picture, and announce "Up" to the grader.

(6) When the tower operator gives the command FIRE, engage the target given in the fire command until you hit it. Fire any remaining rounds at the second target. You need no other fire command. Before firing, you must know the procedure to follow in the event of a stoppage. For each round you fire, your assistant announces "Hit" or "Miss."

e. **Station 3, Task 4, Midrange Supported Position, Point Target, CBRN**.

(1) Put on, clear, and check your mask within 9 seconds. Within the next 6 seconds, pull the hood over your head and zip the front of it closed.

(2) Load one of the three rounds allotted.

(3) When you receive the following fire command, repeat each element as it is given:

FIRE MISSION
FRONT
3 ROUNDS
170 METERS (BUNKER)
AT MY COMMAND

(4) Acquire the proper sight picture and announce "Up" to your assistant.

(5) Have your assistant signal the tower operator that you are ready.

(6) When the tower operator gives the command to commence firing, engage the target given in the fire command until you hit it. For each round you fire, your assistant announces "Hit" or "Miss."

f. **Station 3, Task 5, Midrange Supported Position, Area Target, CBRN**.

(1) Load one of the three rounds allotted.

(2) When you receive the following fire command, repeat each element as it is given:

FIRE MISSION
FRONT
3 ROUNDS
190 METERS (AUTOMATIC WEAPON POSITION)
AT MY COMMAND

(3) Acquire the proper sight picture and announce "Up" to your assistant.

(4) Have your assistant signal the tower operator that you are ready.

(5) When the tower operator gives the command to FIRE, engage the target given in the fire command until you hit it. For each round you fire, your assistant announces "Hit" or "Miss."

5-12. DAY RECORD FIRE QUALIFICATION STANDARDS

Before qualification firing, each grenadier must know the tasks, the time and ammunition required, the procedures to follow if a stoppage occurs, the penalties for failure to stop firing when commanded or signaled to do so, and the method used for scoring targets.

a. **Time and Ammunition**. Each grenadier determines the target and its distance before loading any rounds. When the grenadier receives the command to FIRE, the time allotted for that task in Firing Table I begins.

b. **Stoppages**. The grenadier must apply immediate action procedures if a stoppage occurs. If he can reduce the stoppage, he can continue to fire the course. The trainers allow the grenadier an extra 15 seconds for each application of immediate action.

(1) If a stoppage occurs that you cannot reduce by immediate action, raise your hand and announce "Time."

(2) When you say "Time," the assistant trainer notes the time, ensures that a real stoppage exists, and tries to clear the stoppage. If he clears it, you can complete firing. If he cannot clear it, the grader will clear it, and you will be allowed 15 seconds for each round remaining to complete firing.

(3) If you made an error that caused the stoppage, you do not receive extra time, and your score consists only of whatever you had earned when the stoppage occurred.

(4) If the grenade launcher must be replaced, you are allotted 10 rounds to zero a new one; then you may repeat the exercise.

(5) If malfunctions prevent you from finishing the exercise in the time allowed, you can finish it in an "alibi run" after all other grenadiers complete firing.

c. **Penalties**. Five points are deducted from the score of any grenadier who fails to stop firing when the trainer commands or signals to do so. If a grenadier fires at the wrong target, he loses the rounds allotted for the other target, which leaves him only the remainder of his rounds to expend on both targets.

d. **Target Scoring**. The trainer or assistant trainer records scores on DA Form 2946-R. They determine whether each grenade fired is a hit or miss and assign 0 points for a miss or 10 points for a hit. Tasks 1 through 3 each consist of two targets, so the total available for each of these tasks is 20 points. The grenadier may select which of the two targets to engage first. If he scores a hit on the first, the trainer permits him to engage the second. Once he hits both targets, he returns any unexpended rounds to the assistant trainer. Tasks 4 and 5 each consist of firing one target for a total of 10 points each.

(1) ***Window or Door***. To score a hit on a window or door, the grenade must either strike the target or go through the opening in the center of the target.

(2) ***Bunker***. To score a hit on a bunker, the grenade must strike anywhere on the face of the bunker.

(3) ***Automatic Weapon***. To score a hit on an automatic weapon, the grenade must strike within 5 meters of the target.

(4) ***Troops***. To score a hit on troops, the grenade must strike within 5 meters of the target.

5-13. MOUNTING OF AN/PVS-4 (WITHOUT THE RAIL SYSTEM)

The grenadier must mount the AN/PVS-4 to the weapon before he zeroes it, and he must do both before he can qualify with the M203 grenade launcher. To mount the scope, the grenadier must--

a. Remove the quadrant sight.

b. Position the mounting bracket assembly on the left side of the rifle so that the two clamps project through the opening under the handle. Loosen the wing nuts completely (Figure 5-16, page 5-24).

c. Turn the clamp plates so that the pointed ends are in the UP position and are seated against the handle.

d. Tighten the wing nuts clockwise until the mounting bracket is secure against the weapon.

e. Position the sight in the groove on top of the bracket and align the threaded hole in the base of the sight mounting adapter with the lever screw assembly. Tighten the screw clockwise firmly to secure the sight to the bracket.

5-14. ZEROING OF AN/PVS-4 TO M203

After being mounted on the M203, the nightsight must be zeroed to the M16 before it can be zeroed to the M203. FM 3-22.9 provides instructions for doing this. The M16 is then used to zero the nightsight to the M203. The grenade launcher rounds are fired only to confirm the zero. To zero the nightsight to the M203--

a. **Center the Reticle Pattern**. Use the aiming points on the nightsight reticle (Figure 5-17, page 5-24) and the range settings on the mounting bracket. Center the nightsight's reticle pattern within the field of view (FOV). Note that it may not be centered even if it appears to be. To ensure it is, rotate the azimuth control knob either way until it stops. Then, rotate it back the opposite way, counting the number of clicks until it stops again (this may be any number of clicks between 200 and 600). Divide the number of clicks in half and rotate the knob in the original direction by that number of clicks. For example, if the total number of clicks is 500, rotate the knob back 250 clicks in the original direction. Center the elevation using the same procedure with the elevation control knob. The total amount of elevation clicks also varies between 200 and 600.

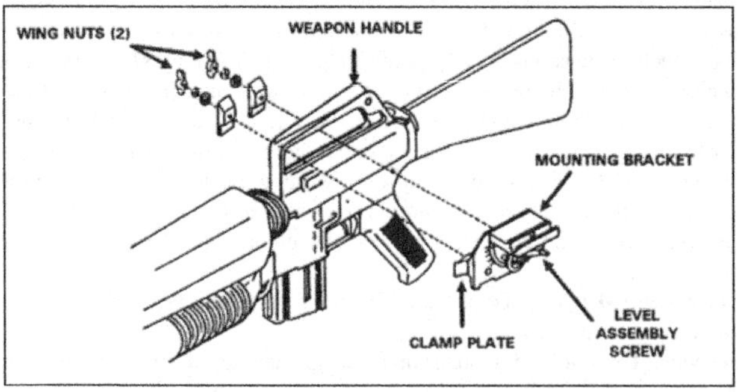

Figure 5-16. Installing the mounting bracket.

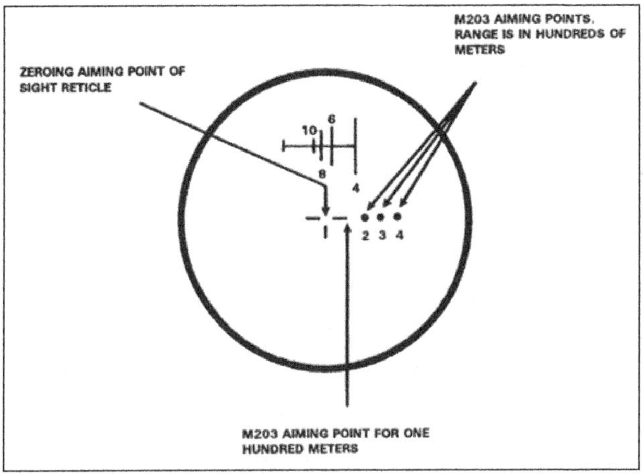

Figure 5-17. Aiming points.

b. **Adjust the Reticle Pattern**. Before adjusting the reticle pattern, the grenadier should fire three 5.56-mm rounds and then retighten the mount wing nuts to securely seat the sight. Once this is done, the grenadier fires at a 10-meter target because hitting and spotting this target is easier than hitting a 25-meter target. This procedure may be performed in daylight using the daylight cover:

(1) Turn the sight on and adjust the reticle intensity to the desired level of illumination.

(2) Place an M16 25-meter target at 10 meters and stabilize the weapon.

(3) Fire a 5.56-mm round at the center of the target and mark the hole the round makes.

(a) If the round misses the entire target, reseat the sight exactly as previously described.

(b) If the round hits the target but not within 20 centimeters (8 inches) of the center, adjust the azimuth and elevation controls to bring the impact point toward the center of the target. Then fire a second round. Continue to fire single rounds and adjust the controls until the rounds strike within the desired distance from the center.

(4) Once the reticle is adjusted, move the 25-meter target out to 25 meters and zero the grenade launcher. Do *not* remove the nightsight from the weapon until you have obtained a zero.

c. **Zero at 25 Meters**. This zero is not recorded. To obtain a 25-meter zero, the grenadier must—

(1) Stabilize the weapon.

(2) Center the reticle's zeroing range aiming point on the target aiming point, which is in the center of the target (Figure 5-18). Fire until you obtain a good three-round shot group. Triangulate and locate the center of the shot group.

NOTE: Even if the nightsight is dismounted and remounted on the same weapon, some changes in its zeroing will occur, so it must be zeroed again.

(3) Turn the azimuth and elevation control knobs to adjust the sight reticle. Move the center of the shot group 9.8 centimeters (3 7/8 inches) below and 4.2 centimeters (1 5/8 inches) to the right of the target aiming point (Figure 5-15). For example, if the shot group is high and to the left of the desired impact point, adjust the elevation down (DN) and the azimuth right (RT). One click of the azimuth or elevation adjustment moves the strike of the round .63 centimeter (1/4 inch) at a range of 25 meters. Two clicks move the reticle about one square on the target.

(4) After adjusting the reticle, assume a stable position. Place the reticle aiming point on the target aiming point and fire three more rounds.

(5) Repeat steps 3 and 4 until the rounds strike within a 3.2 centimeter (1 1/4 inch) circle in the desired location 9.8 centimeters (3 7/8 inches) below and 4.2 centimeters (1 5/8 inches) to the right of the aiming point, or until you have fired 12 rounds, whichever occurs first. If you are unable to zero the AN/PVS-4 after 12 rounds, the trainer must send you to remedial training.

(6) Confirm the zero on the grenade launcher range using a 200-meter target. Place the nightsight into operation and use its reticle, which has two parts. Use the vertical line in the upper part of the reticle to estimate range and the lower part to aim the weapon.

(a) Set the range as estimated on the range indicator of the mounting bracket (Figure 5-16).

(b) Engage the target, placing the aiming point of the sight reticle on the target's center of mass (Figure 5-18). Fire the weapon using all your marksmanship skills. You have confirmed the zero if two of three rounds strike within 5 meters of the target.

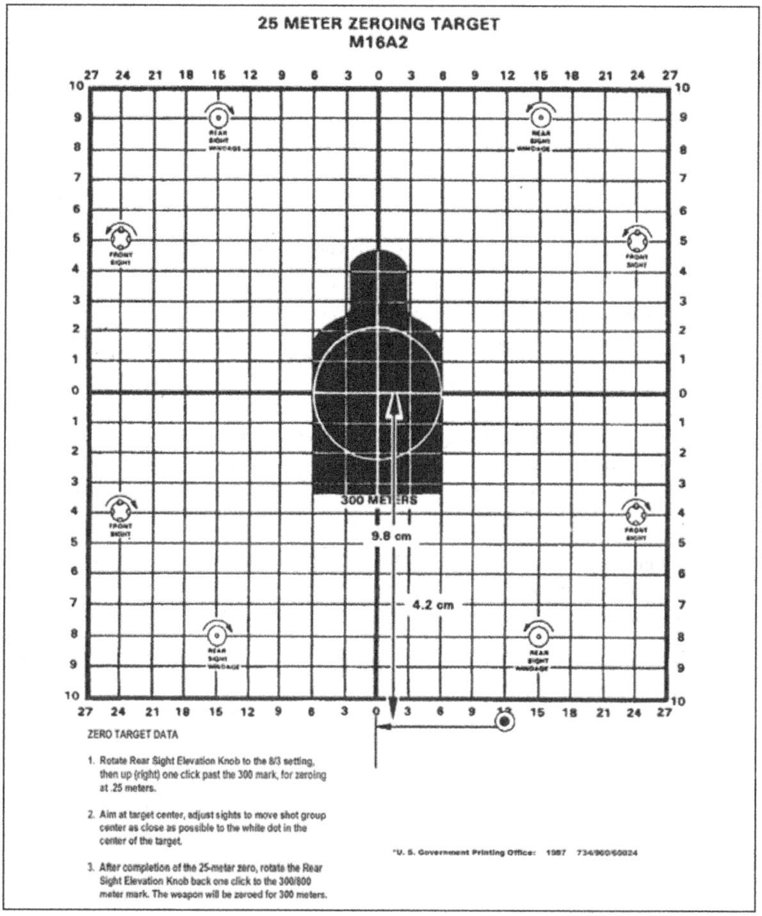

Figure 5-18. Adjustment of rounds.

5-15. NIGHT RECORD FIRE

Night or limited visibility firing trains grenadiers to apply the fundamentals of grenade launcher marksmanship while using the AN/PVS-4 nightsight. This training increases the grenadiers' confidence. Each grenadier learns to zero the M203 to the AN/PVS-4 on the 25-meter range and then fires at one area target on the M203 grenade launcher range. Before night firing, grenadiers receive instruction in its objectives, fundamentals, fire commands, and targets. The unit is organized in firing orders, each consisting of a grenadier and assistant, based on the range constraints. The assistant performs his duties in a manner similar to day record fire. Night record fire consists of one task, *Station 3, Task 6, Mid-Range Fighting Position, Area Target, Night*:

a. Load one of the three rounds allotted.

b. When you receive the following fire command, repeat each element as it is given:

GRENADIER
FRONT
3 ROUNDS
190 METERS (AUTOMATIC WEAPON POSITION)
AT MY COMMAND

c. Acquire the proper sight picture and announce "Up" to the grader.

d. When the grader gives the command FIRE, engage the target given in the fire command until you hit it. Fire any remaining rounds at the second target. You need no other fire command. For each round you fire, your assistant announces "Hit" or "Miss."

5-16. NIGHT RECORD FIRE QUALIFICATION STANDARDS

Before qualification firing, each grenadier must know the task, the time and ammunition required for each, the procedures to follow if a stoppage occurs, the penalties for failure to stop firing when commanded or signaled to do so, and the method used for scoring targets.

a. **Time and Ammunition**. Firing Table II (Table 5-2) provides the night firing task and its time and ammunition requirements.

b. **Stoppages**. The procedure for stoppages is the same as for other qualification firing exercises.

c. **Penalties**. The procedure for penalties is the same as for other qualification firing exercises.

d. **Target Scoring**. The target scoring procedure is the same as for other qualification firing exercises. Record the score in the appropriate block on DA Form 2946-R.

e. **Conditions**. Night record fire trains the grenadier to engage targets between 165 and 250 meters under ideal moonlight conditions.

STATION	TASK	TIME	ROUNDS	TYPE	TARGET(S) AND RANGE(S)
3	3	2 min	3	HE	Automatic weapon at 190 to 200 meters

Table 5-2. Firing Table II, night record fire qualification.

This page intentionally left blank.

(3) ***Linear Targets with Depth***. The grenadier lays on the target's center of mass. He then moves the grenade launcher left and right across the target, selecting successive aiming points at different ranges (Figure 6-9).

Figure 6-9. Linear targets with depth.

* 6-6. DECONTAMINATION

Leaders must try to reduce the penetration of contaminants and lessen exposure to them. Contaminated material is disposed of in accordance with the SOP. Refer to Chapter 2 for decontamination procedures.

Section II. FIRE CONTROL

Fire control includes all leader and Soldier actions in planning, preparing, and applying fire on a target. The leader selects and designates targets, indicating their width and depth or, in the case of targets that are hard to identify, designates the distance from a reference point to the target's center of mass (Figure 6-10, page 6-10). He also designates the midpoint, flanks, or ends of a target unless these locations are obvious to the grenadiers. The grenadiers open fire when ready, adjust and regulate the rate of fire, and shift from one target to another. They cease fire only when the target is neutralized or the leader signals to cease fire.

Figure 6-10. Use of a reference point to identify a target.

6-7. METHODS OF FIRE CONTROL

The noise and confusion of battle may limit the methods of fire control used, so the leader must select the method(s) that will best accomplish the mission.

a. **Oral Commands**. The primary method of fire control is the oral fire command. This method is effective unless noise or distance prevents the grenadier from hearing the leader.

b. **Arm-and-Hand Signals**. This method of fire control is effective only if the grenadiers know the standard arm-and-hand signals and can see the leader. The leader gets the grenadier's attention, then points to the target. When the grenadier returns the "ready" signal, the leader commands the grenadier to fire.

c. **Prearranged Signals**. This method of fire control can include visual or sound signals such as those that can be produced by a whistle, pyrotechnics, or casualty-producing device. The SOP must define the signals to be used, and all squad members must understand them. If the leader wants to shift fire at a certain time, he gives a prearranged signal such as smoke or pyrotechnics. When they see this signal, grenadiers shift their fire to a prearranged point.

d. **Personal Contact**. This method of fire control is the one most frequently used by leaders of small units. Many situations require the leader to move to individual Soldiers to issue orders. If so, he must use cover and concealment to avoid disclosing their positions. Once there, he gets the grenadier's attention, points out the new target, and commands FIRE.

e. **Standing Operating Procedure**. This method of fire control refers to actions executed without command. The SOP defines these actions and the events that initiate them. Using an SOP simplifies the leader's job of fire control.

f. **Range Cards**. This method of fire control requires the leader to ensure all range cards are current and accurate. He designates dead spaces, specific targets, no-fire zones, and restricted fire areas. The key to this method is the disciplined grenadier who pays attention to detail and can understand the areas the squad leader wants covered by fire. Figure 6-11 shows an example of an M203 range card.

*GLOSSARY

AAR	after-action review
ammo	ammunition
C	Centigrade
CBRN	chemical, biological, radiological, and nuclear
CCC	Captain's Career Course
CLP	cleaner, lubricant, preservative
cm	centimeter(s)
CMF	career management field
CS	a chemical agent (tear gas)
DA	Department of the Army
DODAC	Department of Defense Ammunition Code
EST	engagement skills trainer
F	Fahrenheit
FM	field manual
FOV	field of view
FPL	final protective line
fps	feet per second
FTX	field training exercise
HE	high explosive
HEDP	high-explosive dual-purpose
hr	hour(s)
IAW	in accordance with
IET	initial entry training
KD	known-distance (range)
kg	kilogram(s)
kps	kilometers per second
LFX	live-fire exercise
LTA	local training area
METL	mission-essential task list
m	meter(s)
min	minute(s)
mm	millimeter(s)
MOPP	mission-oriented protective posture
mph	mile(s) per hour

mps	meter(s) per second
MTA	major training area
MTP	mission training plan
NCO	noncommissioned officer
NCOES	Noncommissioned Officer Education System
NCOIC	noncommissioned officer in charge
NSN	national stock number
NVD	night vision device
OBC	Officer Basic Course
OIC	officer in charge (of)
OGIVE	head of the round
PD	point-detonating
PDF	principal direction of fire
PMT	preparatory marksmanship training
psi	pounds per square inch
RETS	remoted target system
rpm	revolution(s) per minute
S2	Intelligence Officer
S3	Operations and Training Officer
S4	Supply Officer
SEE	small emplacement excavator
SM	soldier's manual
SMCT	soldier's manual of common tasks
SOP	standing operating procedure
SSN	social security number
STP	soldier training publication
STX	situational training exercise
TP	training practice
TRADOC	Army Training and Doctrine Command
TRC	training readiness condition
UO	urban operations

*REFERENCES

SOURCES USED

These are the sources quoted or paraphrased in this publication. All are US Army publications unless stated otherwise.

FM 3-3 Chemical and Biological Contamination Avoidance. 16 November 1992, with Change 1, 29 September 1994.

FM 3-11.4 Multiservice TTP for NBC Protection. 2 June 2003.

FM 3-11.5 Multiservice TTP for Chemical, Biological, Radiological, and Nuclear Decontamination. 4 April 2006.

FM 21-60 Visual Signals. 30 September 1987.

FM 21-75 Combat Skills of the Soldier. 3 August 1984.

FM 3-22.9 Rifle Marksmanship, M16A1, M16A2/3, M16A4, and M4 Carbine. 24 April 2003, with Changes 1 through 3, 22 January 2004 through 28 April 2005.

STP 7-11B1-SM-TG Soldier's Manual and Trainer's Guide, MOS 11B, Infantry, Skill Level 1. 28 May 2004.

STP 7-11B24-SM-TG Soldier's Manual and Trainer's Guide, MOS 11B, Infantry, Skill Levels 2, 3, and 4. 28 May 2004.

STP 21-1-SMCT Soldier's Manual of Common Tasks, Skill Level 1. 11 October 2005.

TC 25-8 Training Ranges, 25 February 1992, with Change 1, 5 April 2005.

TM 9-1010-221-10 Operator's Manual for 40-mm Grenade Launcher, M203. 1 August 2001.

TM 9-1010-221-23&P Unit and Direct Support Maintenance Manual, Including Repair Parts and Special Tools List, for Launcher, Grenade, 40-mm, M203. 12 January 1993.

TM 43-0001-28 Army Ammunition Data Sheets for Artillery Ammunition: Guns, Howitzers, Mortars, Recoilless Rifles, Grenade Launchers, and Artillery Fuzes. 28 April 1994 with Changes 1 through 13, 30 May 1995 through 31 March 2005.

TM 750-244-7	Procedures for Destruction of Equipment In Federal Supply Classifications 1000, 1005, 1010, 1015, 1020, 1025, 1030, 1055, 1090, and 1095 To Prevent Enemy Use. 18 June 1970, with Change 1, 29 October 1970.

DOCUMENTS NEEDED

These documents must be available to the intended users of this publication.

AR 385-63	Policies and Procedures for Firing Ammunition for Training, Target Practice, and Combat. 19 May 2003.
DA Form 2946-R	40-mm Grenade Launcher Scorecard.
DA Pam 350-38	Standards in Weapons Training. April 2006.
TRADOC Reg 350-6	Enlisted Initial Entry Training (IET) Policies and Administration. 30 December 2005.

WEB SITES

Army Knowledge Online	https://akocomm.us.army.mil/usapa/doctrine/index.html
Army Publishing Directorate (for forms)	http://www.apd.army.mil/usapa_PUB_formrange_f.asp
Reimer Digital Library	http://www.train.army.mil

*INDEX

40-MM GRENADE LAUNCHER SCORECARD

For use of this form, see FM 3-22.31; the proponent agency is TRADOC.

1. ID CODE *	2. UNIT	3. DATE *(YYYYMMDD)*

4. ZERO LEAF SIGHT	DEFILADE	ELEVATION	5. ZERO QUADRANT SIGHT	DEFILADE	ELEVATION

6. RECORD FIRE	TASK NO.	TIME (MIN)	TGT	HIT	MISS
DAY	1	2	1		
			2		
	2	2	1		
			2		
	3	2	1		
			2		
CBRN	4	2	1		
	5	2	1		
NIGHT	6	2	1		
* Do not use personal information	7. TOTAL POINTS *(Award 10 points for each hit)*				

8. RATING SCALE

TOTAL POINTS *(Block 7)*	RATING
80 - 90	EXPERT ☐
70 - 75	FIRST CLASS ☐
60 - 65	SECOND CLASS ☐
0 - 55	UNQUALIFIED ☐

9. GRADER'S INITIALS	10. DATE *(YYYYMMDD)*
11. OIC'S INITIALS	12. DATE *(YYYYMMDD)*

DA FORM 2946-R, FEB 2007 EDITION OF NOV 2002 IS OBSOLETE. APD V1.00

FIELD MANUAL
NO. 3-22.27

HEADQUARTERS
DEPARTMENTS OF THE ARMY
WASHINGTON, DC, 28 November 2003

MK 19, 40-mm GRENADE MACHINE GUN, MOD 3

CONTENTS

*This publication supersedes FM 23-27, dated 27 December 1988.

PREFACE

This manual discusses how to train the unit to use the MK 19, 40-mm grenade machine gun, model (MOD) 3, referred to in this manual as the MK 19. This manual highlights mechanical training, weapon capabilities, and gunnery principles, methods, techniques, and standards that apply to the MOD 3. It also includes preliminary gunnery, a gunnery skills test, gunnery tables, and qualification tables. If this information conflicts with an applicable technical manual (TM), revised TM, or TM update, the crew will follow the guidance in the most recently published document.

The primary audiences for this manual are soldiers, trainers, and staff officers. Units can modify the gunnery program to meet local training restraints. Although the gunnery tables are intended for use with live fire, trainers can use the tactical engagement simulator system (TESS), a video disc trainer (VDT), a multipurpose arcade combat simulator (MACS), or other training device, except on qualification tables.

In all cases, units must evaluate their training to ensure that it follows the building-block principle and adheres to sound training policy. Only a crew that is trained and does well in preliminary gunnery exercises is likely to do well in live-fire exercises (LFX) and in combat situations.

The proponent for this publication is the U.S. Army Infantry School. Send comments and recommendations to doctrine@benning.army.mil or on a DA Form 2028 (Recommended Changes to Publications and Blank Forms) directly to Commandant, U.S. Army Infantry School, ATTN: ATSH-ATD, Fort Benning, Georgia 31905-5593.

Unless this manual states otherwise, masculine nouns and pronouns do not refer exclusively to men.

CHAPTER 1
INTRODUCTION

Although the MK 19 is a recent entry into the Army's inventory, development began in 1963. The first version was a hand-cranked, multiple grenade launcher called the MK 18. In 1966 the need for more firepower inspired the development of a self-powered 40-mm machine gun called the MK 19, MOD 0. This model was neither reliable nor safe enough for use as a military weapon system. Product improvements begun in 1971 resulted in the 1972 MOD 1, of which only six were produced. The MOD 1 performed effectively in Navy riverine patrol craft and broader applications for the MK 19 were found. In 1973 the Navy developed the MOD 2, which featured improved reliability, safety, and maintainability. In 1976 a complete redesign resulted in the MK 19, MOD 3, which the Army adopted in 1983. The Army now uses the MK 19 within the tactical environment for defense, retrograde, patrolling, rear area security, urban operations, and special operations.

This chapter provides applications, training strategies, and descriptive, technical, and operational data for the MK 19 (Figure 1-1).

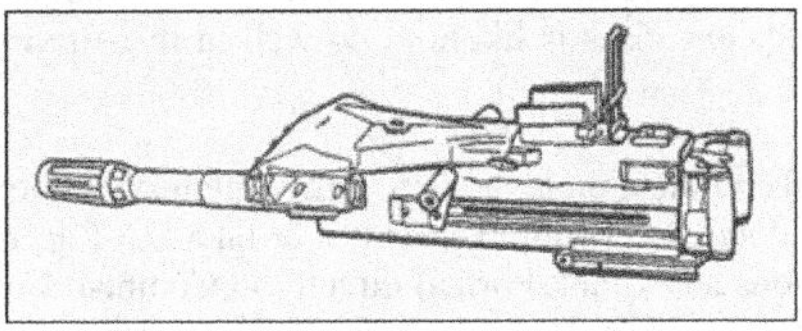

Figure 1-1. MK 19, 40-mm grenade machine gun, MOD 3.

1-1. APPLICATIONS

The MK 19 supports the soldier in both the offense and defense. It gives the unit the capability of laying down a heavy volume of close, accurate, and continuous fire. The MK 19 can also:

- Protect motor movements, assembly areas, and supply trains in a bivouac.
- Defend against hovering rotary aircraft.
- Destroy lightly-armored vehicles.
- Fire on suspected enemy positions.
- Provide high volumes of fire into an engagement area (EA).
- Cover obstacles.
- Provide indirect fires from defilade positions.

1-2. DESCRIPTION

The MK 19 is an air-cooled, blowback-operated machine gun with five major assemblies (Figure 1-2). A disintegrating metallic link belt feeds ammunition through the left side of the weapon. Tables 1-1 and 1-2 provide MK 19 technical and operational data, respectively.

a. **Receiver Assembly.** Holds the barrel and other parts of the gun. Ammunition is fed into the left side of the receiver through the feed throat assembly. The MK 19's barrel will not overheat, even after prolonged firing.

b. **Feed Slide Assembly and Tray.** Holds the rounds in the feeder and indexes the ammunition into position for delinking.

c. **Top Cover Assembly.** Holds the feed slide assembly and tray. It is opened by a latch (left side) for loading or to clean and inspect feeder area. A blade-type front sight is attached to the top cover assembly (Figure 1-3).

d. **Sear Assembly.** Holds the receiver sear. Trigger action releases the sear and allows the bolt to go forward. The safety is attached to the sear assembly.

e. **Bolt and Backplate Assembly.** The bolt fires the round when the sear is depressed by trigger action. The recoil springs drive the bolt forward on the receiver rails. The guide rods hold the springs in position. Trigger and handgrips are located on the backplate assembly.

f. **Feed Throat Assembly.** Allows smooth feeding of 40-mm ammunition. It attaches to the forward left side of the receiver by two sets of spring-loaded retaining pins. Without a feed throat, machine gun stoppages may occur because of twisted or misaligned rounds.

g. **Leaf-Type Rear Sight (with adjustable range plate).** Is marked in 100-meter intervals from 300 to 1,500 meters. The sight is mounted on a spring dovetail base to the receiver assembly (Figure 1-4). Before moving the weapon, the gunner folds the sight forward to a horizontal position. The rear sight can be adjusted for range and windage.

(1) ***Range.*** Different adjustments can be made to the range. Use the rear sight slide release to make *major adjustments* to the range. Use the elevation wheel to make *fine adjustments* to the range.

(2) ***Windage.*** Use the rear sight to adjust for windage. One click equals a 1-mil change. To move the sight to the *right*, turn the windage screw clockwise. To move the sight to the *left*, turn the windage screw counterclockwise.

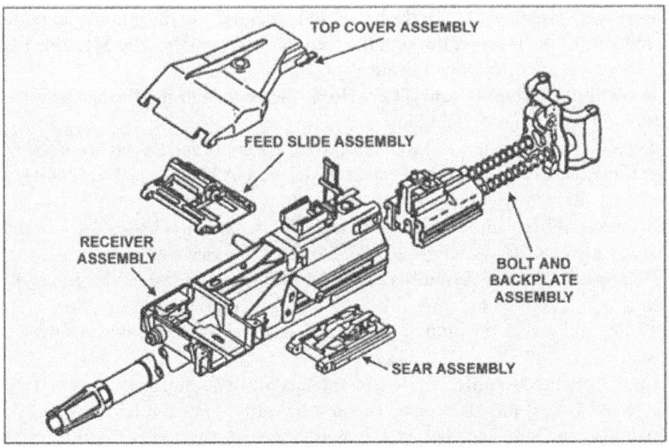

Figure 1-2. Five major assemblies.

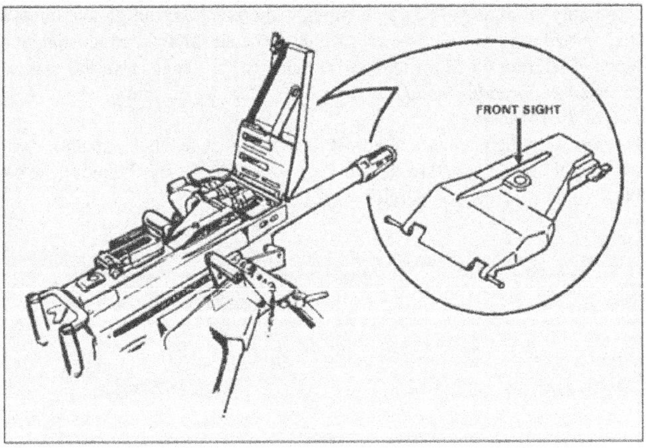

Figure 1-3. Front sight on top cover assembly.

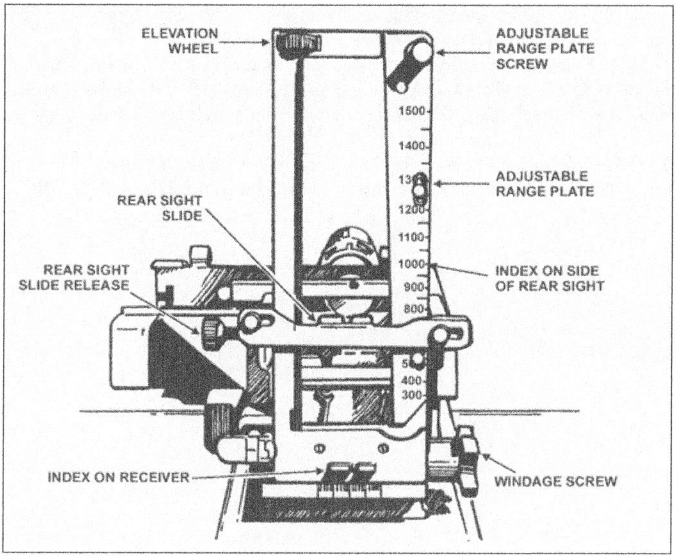

Figure 1-4. Rear sight.

> **WARNING**
>
> **Use only prescribed ammunition. Mixing types of ammunition could result in injury.**

h. The MK 19 uses the following 40-mm cartridges (Figure 1-5, page 1-8).

(1) ***High-Explosive, Dual-Purpose M430 Cartridge.*** The high-explosive, dual-purpose (HEDP) M430 cartridge is the standard round for the MK 19 (Department of Defense Identification Code [DODIC] B542). They are linked with M16A2 links. The HEDP round, the top-curved portion of the projectile, is olive drab with a yellow ogive and yellow markings. It is packed in M548 (48 rounds) or PA120 (32 rounds) ammunition containers. The HEDP, an impact-type round, can penetrate 2 inches of steel armor at 0-degree obliquity and inflict personnel casualties out to 15 meters from impact. It arms within 18 to 30 meters of the gun muzzle and has a point-initiating, base-detonating (PIBD) fuze.

(2) ***High-Explosive M383 Cartridge.*** The high-explosive (HE) M383 cartridge round is olive drab with a yellow ogive and yellow markings. It is packed in a metal ammunition container (48 rounds, linked, in each container). The HE round has a wound radius of 15

meters. It lacks the armor-penetrating ability of the HEDP M430 round. The HE arms between 18 to 36 meters of the gun muzzle fuze.

(3) ***M922 Dummy Cartridges.*** Each MK 19 is issued with one 10-round of inert dummy rounds belt (DODIC B472). M16A2 links join the dummy rounds into a 10-round belt packed in an M2A1 metal box. Trainers use dummy rounds to check weapon function and to train crews.

(4) ***M918 Cartridge.*** The M918 is a training practice cartridge that has the same muzzle velocity of 790 feet per second (fps), signature, and sound as the HE round (DODIC B584).

MK 19 TECHNICAL DATA		
MK 19 (MOD 3):		
Weight without feed throat	77.6	pounds
Weight with feed throat	78.0	pounds
Length	43.1	inches
Width	14.0	inches
Height	8.8	inches
MK 64 (MOD 7) gun cradle:		
Weight	21.0	pounds
Length	17.5	inches
Height	9.5	inches
Tripod (M3) weight:	44.00	pounds
Gun and cradle:		
Weight without feed throat	98.6	pounds
Weight with feed throat	99.0	pounds
Gun, cradle, and tripod:		
Weight without feed throat	142.6	pounds
Weight with feed throat	143.0	pounds
Mounts:	**Ammunition:**	
M3 tripod	M430 (HEDP)	
M4 pedestal	M383 (HE)	
M66 ring	M918 (TP)	
HMMWV weapon platform	M922 (dummy)	
M113 APC commander's cupola		
NOTE: Keep the feed throat attached to the weapon.		

Table 1-1. Technical data.

MK 19 OPERATIONAL DATA		
Maximum range:	2,212	meters
Maximum effective range:	1,500	meters (point target)
	2,212	meters (area target)
Rates of fire:		
Sustained	40	rounds per minute
Rapid	60	rounds per minute
Cyclic	325 to 375	rounds per minute
Ammunition:		
M430 HEDP	2	inch armor
	15	meter casualty radius
M383 HE	15	meter casualty radius
Service frequency:	50,000	rounds
Elevation, tripod controlled:	100	mils
Depression, tripod controlled:	258	mils
Traverse, tripod controlled:	800	mils (400 left plus 400 right)
Muzzle velocity (average):	798	feet per second
Recoil forces (average):	500	pounds
Angle of automatic fire:	0 to 70	degrees elevation (automatic fire), based on mounting arrangements
Weights:		
Rounds	62	pounds (48 rounds in M548 metal container)
	42	pounds (32 rounds in PA120 metal container)
Planned operating load	400	prescribed by local commanders

Table 1-2. Operational data.

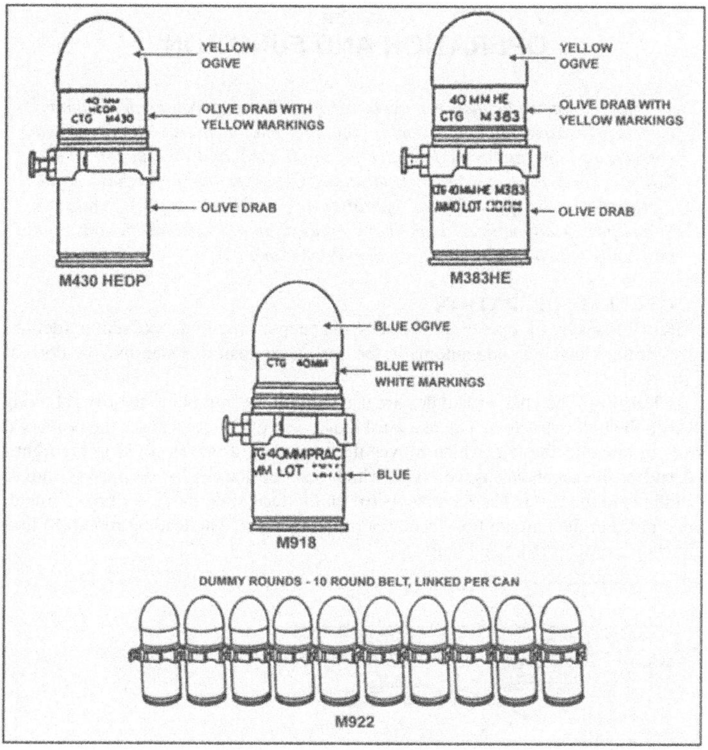

Figure 1-5. 40-mm cartridges.

1-3. TRAINING STRATEGY

A training strategy that integrates resources with the instruction and training of individual and collective skills is necessary to carry out a unit's wartime mission. TRADOC institutions, such as the noncommissioned officers education system (NCOES), and units implement overall training strategies for the MK 19. This multifaceted approach includes specific plans for using resources such as publications, ranges, ammunition, training aids, devices, simulators, and simulations. These specific strategies develop critical soldier skills and training tasks, as well as leader skills needed to support the intended outcome (see Appendix A, Training Strategy, for more details).

CHAPTER 2

OPERATION AND FUNCTION

This chapter includes the cycle of operation; operating precautions; clearing; disassembly and assembly; loading; malfunctions, stoppages, and corrections; operations under adverse conditions; and destruction of the MK 19. Refer to Appendix B, Operations Checklist, for procedures to be used before, during, and after operation of the MK 19 and Appendix C, Stoppages, Malfunctions, Immediate Actions, and Remedial Actions, for procedures to correct stoppages and malfunctions.

2-1. CYCLE OF OPERATION

The MK 19's cycle of operation includes six steps: charging, extracting (delinking), cocking, firing, blowback, and automatic feeding. More than one step may be done at the same time.

a. **Charging.** The charger handles are used to pull the bolt (1) to the rear, aligning the round with the bolt extractors. The rearward movement of the bolt causes the primary drive lever (2) to move to the left, which moves the secondary drive lever (3) to the right. The forked end of the secondary drive lever, which rests on the feed slide pin (4), moves the feed slide (5) to the right. The feed pawls (6) on the feed slide move the linked rounds (7) over one place in the ammunition-feed area of the receiver. The leading round (8) lines up with the bolt extractor (9). See Figure 2-1.

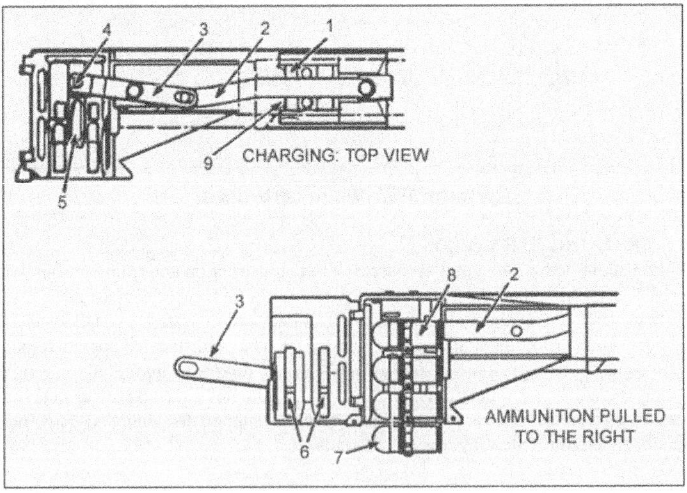

Figure 2-1. Charging cycle.

b. **Extracting (Delinking).** When a round is stripped from the belt, it is extracted or "delinked." This happens, after the MK 19 has been charged once, when the trigger (10) is pressed. The bolt slams forward and the bolt's extractors (11) snap over the rim of the cartridge case. When the MK 19 is charged again, the extractor pulls the leading round to the rear and separates the male and female links. The curved edge of the vertical cam (12) forces the lead round out of the extractors and into the bolt fingers (13). With the bolt completely to the rear, the round lines up with the chamber (14), and is ready to fire. As the original leading round chambers, the next round aligns with the bolt extractors. See Figure 2-2.

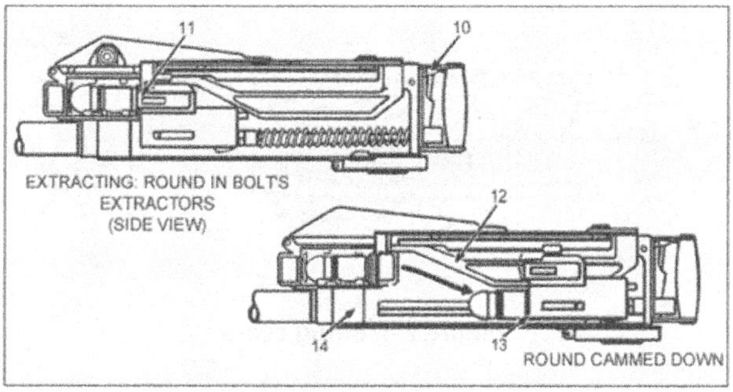

Figure 2-2. Extracting (delinking) round.

c. **Cocking.** The rearward movement of the bolt causes the cocking lever (15) to retract the firing pin (16). When the cocking lever hits the rear end of the left receiver rail slot (17), the cocking lever is forced forward. When the cocking lever retracts the firing pin, the firing pin sear holds the firing pin to the rear (18). See Figure 2-3.

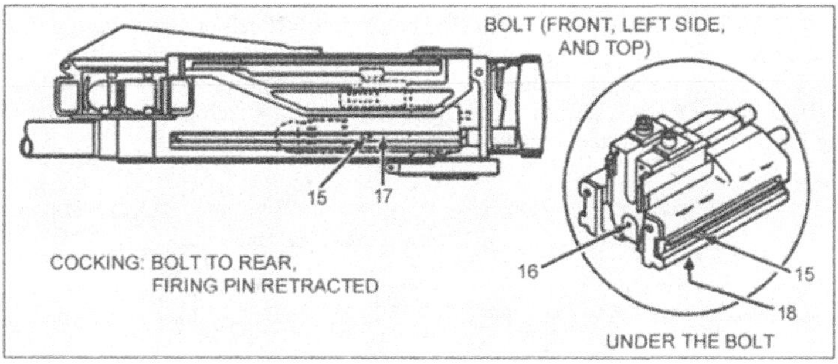

Figure 2-3. Cocking the MK 19.

d. **Firing Cycle.** The safety switch is on F (FIRE). Pressing the trigger (20) depresses the tip of the receiver sear (21). The receiver sear disengages the bolt sear (22), which releases the bolt forward under spring pressure with a round in the bolt fingers. The cocking lever hits the forward end of the left receiver rail slot, forcing the lever to the rear. The bolt sear hits a plate in the bottom of the receiver, which pushes the firing pin sear up to release the firing pin. A combination of the bolt's inertia and pressure from the firing pin spring drive the firing pin forward. The tip of the firing pin detonates the primer. The round is not completely inside the chamber at the moment the weapon is fired. The cartridge case, held by the bolt fingers, protrudes from the chamber (23). The explosion forces the projectile down the bore. See Figure 2-4.

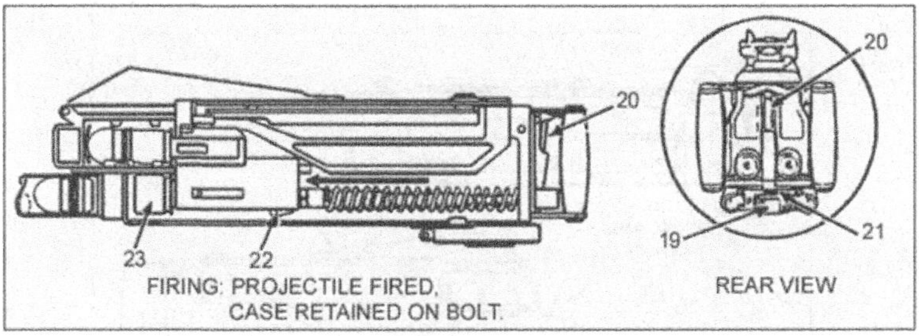

Figure 2-4. Firing cycle.

e. **Blowback and Automatic Feeding Cycle.** The gases from the burning powder force the bolt, with a new round in its extractors, to the rear. During this blowback, several things happen at once. First, the curved rail of the vertical cam delinks and forces down the new round on top of the spent case, which forces the spent case out of the bolt fingers and ejects it out the bottom of the gun. Next, the feed slide assembly pulls the rounds to the right in the receiver ammunition feed area, where a new round is ready to pick up (automatic feed). During the bolt's travel to the rear, the cocking lever is pushed forward, which cocks the firing pin. When the bolt reaches the limit of its rearward travel, the recoil springs (24) are completely compressed. The bolt buffers (25) absorb over-travel, reducing trunnion load (recoil force) at the gun-mount attaching points. The bolt sear will not engage the receiver sear if the trigger is still depressed, and another firing cycle occurs. Release of the trigger causes the bolt sear to engage the receiver sear, which prevents the bolt from going forward, and firing stops. See Figure 2-5.

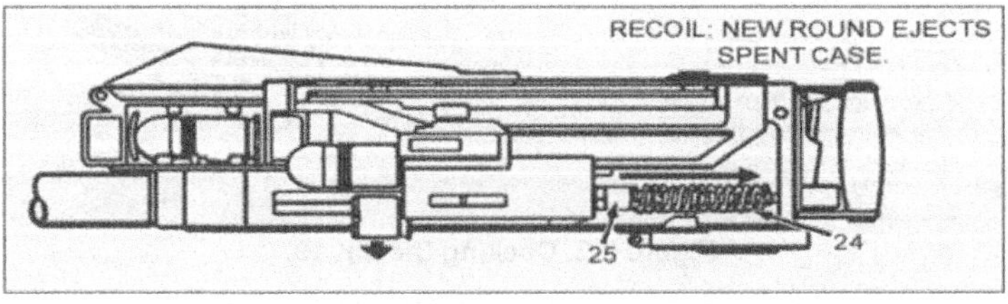

Figure 2-5. Blowback and automatic feed.

2-2. OPERATING PRECAUTIONS

Precautions are taken before, during, and after firing of the MK 19.

a. Before firing the weapon ensure that:

(1) The correct ammunition is on hand.

(2) The ammunition is clean and dent-free.

(3) All ogives on the cartridges are tight.

(4) The line of fire is clear of obstructions.

b. While firing the weapon ensure that:

(1) The top cover is closed.

(2) The muzzle of the MK 19 is pointed downrange at all times.

NOTE: In the event of a runaway gun, lower one of the charging handles.

(3) Use the suggested three to five-round bursts.

c. After firing the weapon:

(1) Unload and clear the MK 19.

(2) Note weapon discrepancies and report them to the armorer.

(3) Clean and lubricate the MK 19 before storage.

2-3. CLEARANCE PROCEDURES

The MK 19 is cleared differently in a firing situation than in a nonfiring situation.

a. **Firing Situation.** In a firing situation, use the following procedures to clear the MK 19:

(1) Move the safety switch to S (SAFE) (Figure 2-6).

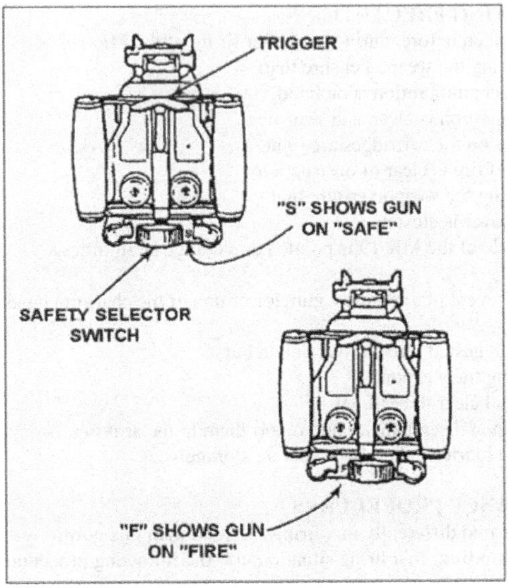

Figure 2-6. S (SAFE) position.

(2) Open the top cover assembly (Figure 2-7). If all the ammunition has NOT been fired, the bolt is to the rear and a round is on the bolt face. If the bolt is forward, lock it to the rear.

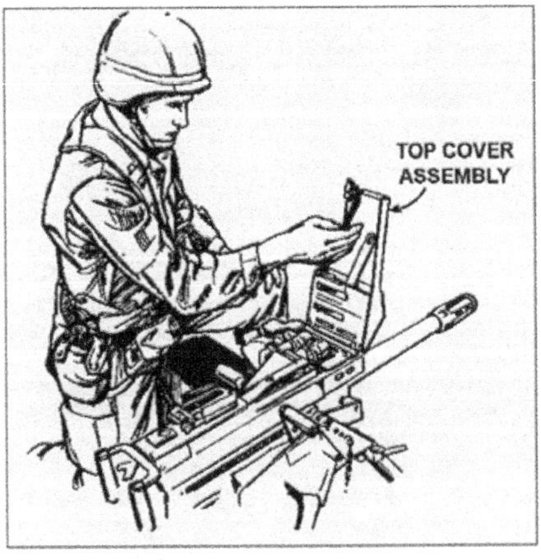

Figure 2-7. Opening the top cover assembly.

(3) Take the ammunition from the feed tray by reaching beneath the feed tray and pressing the primary and secondary positioning pawls. While pressing the position pawls, slide the linked rounds out of the MK 19 through the feed throat (Figure 2-8).

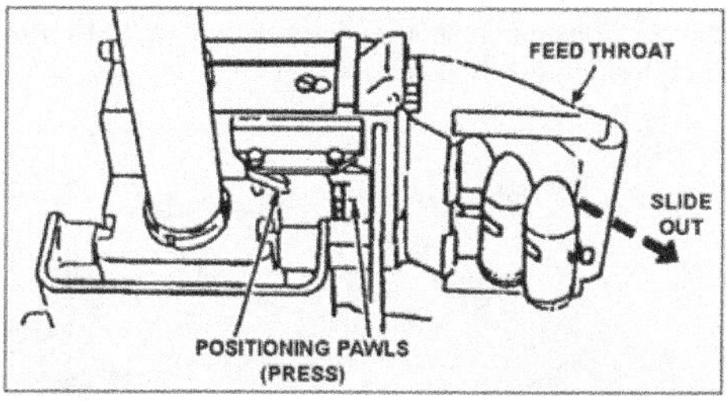

Figure 2-8. Removal of linked rounds from feeder.

(4) Insert a section of the cleaning rod or bayonet through either side of the receiver rail. Place it on top of the live round or cartridge case, as close to the bolt face as possible, and push down. This action forces the round out of the MK 19.

(5) Lower and pull both charger handles to the rear (Figure 2-9).

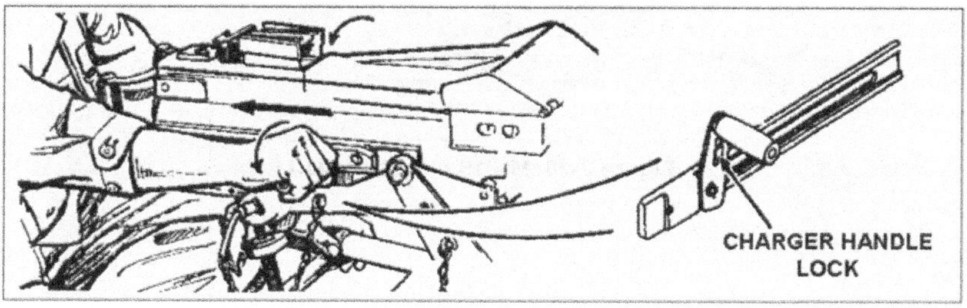

Figure 2-9. Charger handles.

(6) Inspect the chamber and bolt face to ensure that no live rounds are in the weapon.

(7) Place the safety switch on F (FIRE) (Figure 2-6).

(8) Maintain rearward pressure on the charging handle, press the trigger and ease the bolt forward.

(9) Place the safety switch on S (SAFE).

b. **Nonfiring Situation.** In a nonfiring situation, use the following procedures to clear the MK 19:

(1) Place the safety switch on S (SAFE) (Figure 2-6).

(2) Open the top cover assembly.

(3) Lower one or both charger handles.

(4) Pull the charger handle slightly to the rear.

(5) Allow sufficient space between the face of the bolt and the chamber to see both.

(6) Check for live ammunition.

(7) Ride the bolt forward.

(8) Return the charger handle to its original upright position.

2-4. DISASSEMBLY AND ASSEMBLY

Only qualified ordnance personnel should disassemble the MK 19 beyond the steps detailed in this paragraph (Figure 2-10 and Table 2-1).

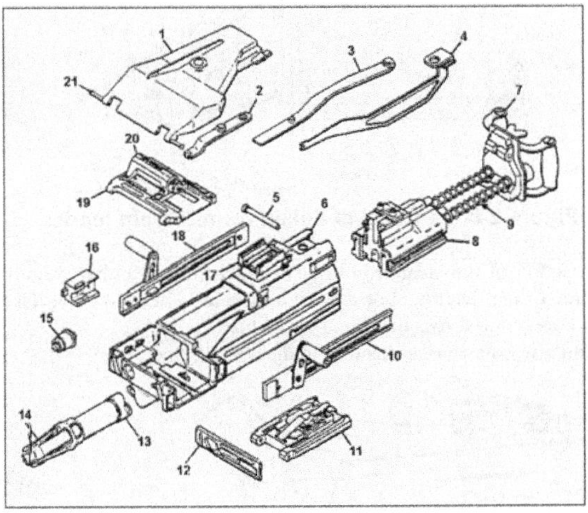

Figure 2-10. MK 19 disassembled.

ITEM	DESCRIPTION
1	Top Cover
2	Secondary Drive Lever
3	Primary Drive Lever
4	Vertical Cam Assembly
5	Backplate Pin
6	Receiver Assembly
7	Control Grip Assembly
8	Bolt
9	Guide Rods and Springs
10	Charger Assembly, Left
11	Sear Assemble
12	Alignment Guide Assembly
13	Barrel
14	Flash Suppressor
15	Ogive Plunger Assembly
16	Round-Positioning Block
17	Rear Sight Assembly
18	Charger Assembly, Right
19	Feed Tray
20	Feed Slide
21	Cover Pins

Table 2-1. Names of the MK 19 major components.

NOTE: High friction areas — the under side of the primary drive lever (3), the vertical cam assembly (4), the bolt (8), the area where the feed slide (20) and feed slide tray (19) touch have been treated to prolong the life of the MK 19 — should be handled carefully during cleaning and handling.

CAUTION

The sear assembly should be removed only in a clean and well-lit area because a small pin could possibly fall out and become lost, which makes the gun inoperable.

a. **Disassembly.** Disassembly includes removal of parts to the extent explained in this chapter. To ensure that parts are not lost and are replaced properly, place them (in the order in which they are taken off) on a clean flat surface. A spent cartridge casing, a section of the cleaning rod, and a cartridge link may be used as removal and replacement tools. To disassemble:

(1) Clear the weapon.

(2) Take out the secondary drive lever by following these procedures:

(a) Raise the top cover assembly and push the secondary drive lever pivot post from the outside of the top cover assembly.
(b) Separate the secondary drive lever from the top cover assembly (Figure 2-11).
(c) Tap lightly on the secondary lever pivot post with any available tool, if necessary.

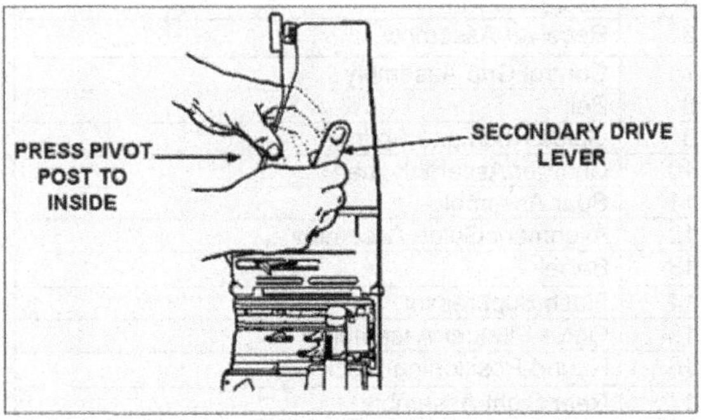

Figure 2-11. Secondary drive lever removal.

(d) Take the secondary drive lever from the slide assembly, and allow the feed slide and tray assembly to close.
(3) Take off the top cover assembly by following these procedures:
(a) Hold the top cover straight up with one hand.
(b) Pull the top cover pins from both sides (Figure 2-12).
(c) Lift the top cover assembly straight up and off.

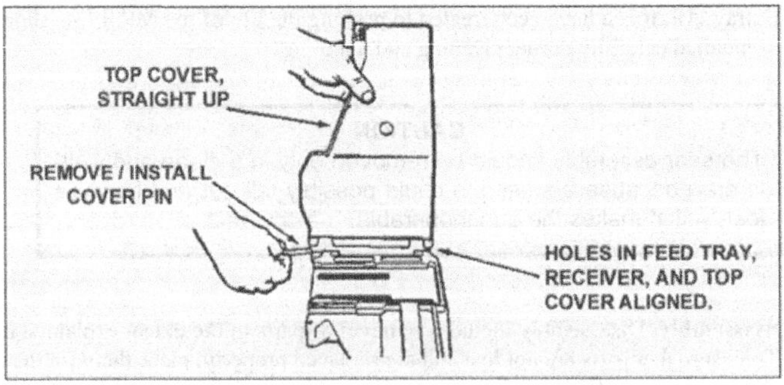

Figure 2-12. Top cover assembly.

(4) Take out the feed slide assembly and separate the feed slide from the feed tray by following these procedures:

(a) Align the tabs on the feed slide with the slots in the feed tray and lift them straight up (Figure 2-13).

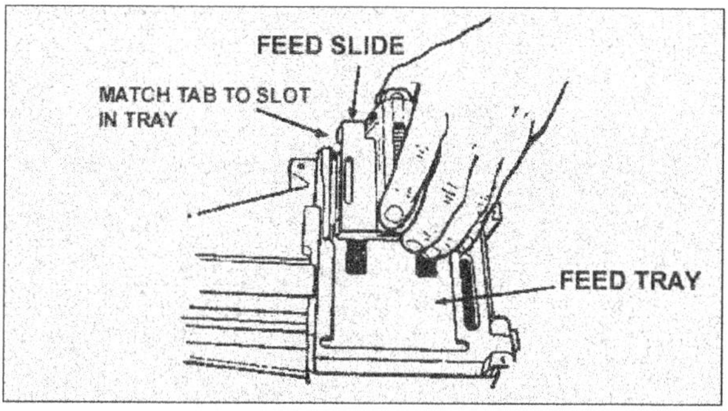

Figure 2-13. Feed slide assembly.

(b) Take out the feed tray by lifting it straight up (Figure 2-14).

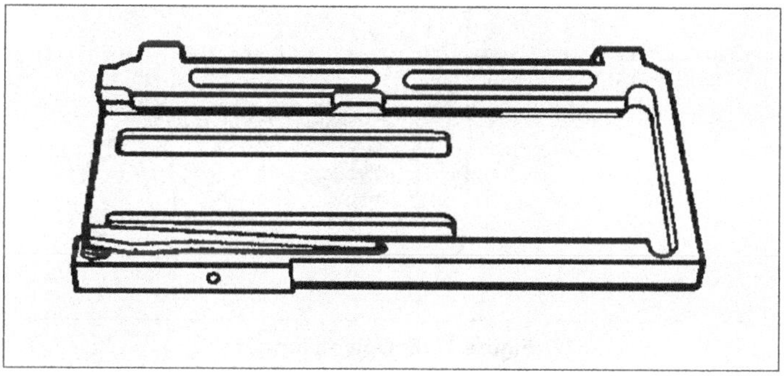

Figure 2-14. Feed tray.

(5) Take off the alignment guide by following these procedures:

(a) Depress the tip of the alignment guide spring with the secondary drive lever or a cleaning rod section.

(b) Slide the alignment guide out of the receiver by pulling the assembly slightly rearward (Figure 2-15).

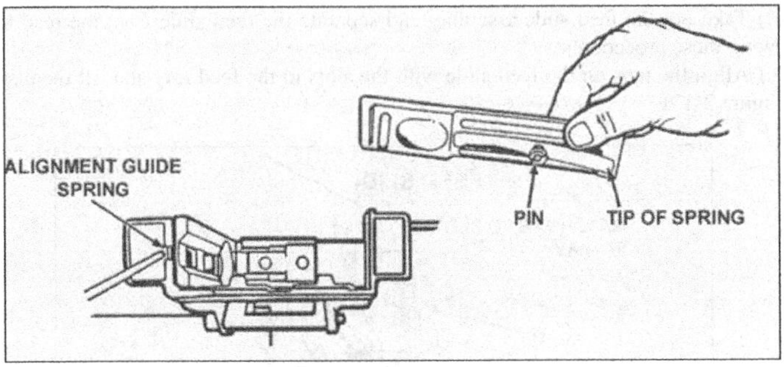

Figure 2-15. Alignment guide.

(6) Take out the ogive plunger by pulling it out through the inside wall of the receiver (Figure 2-16).

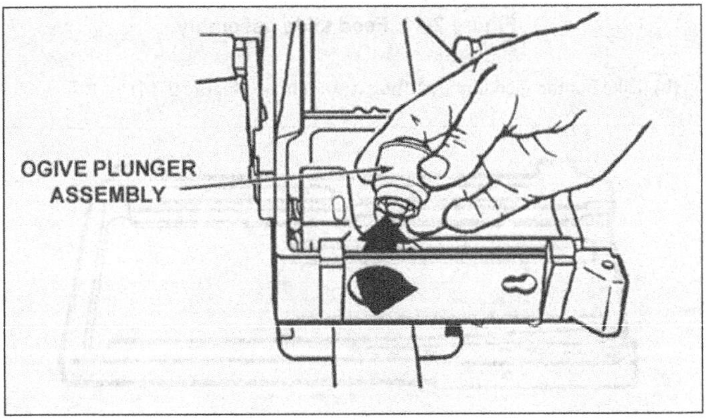

Figure 2-16. Ogive plunger.

(7) Take off the round-positioning block by following these procedures:

(a) Push the round-positioning block into the side of the gun.

(b) Slide the round-positioning block forward.

(c) Release the round-positioning block forward from the keyslots in the receiver wall (Figure 2-17).

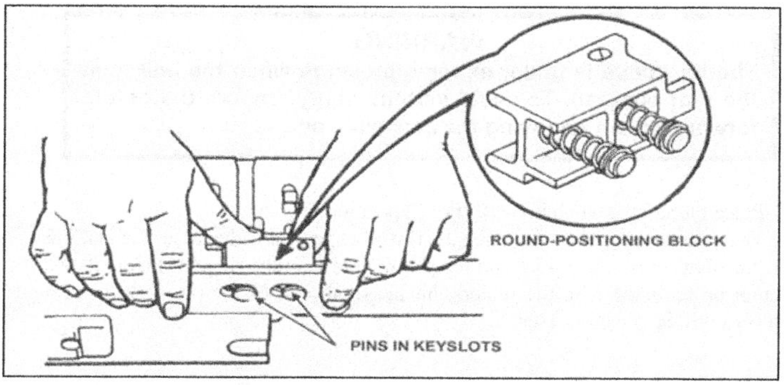

Figure 2-17. Round-positioning block.

(8) Take out the bolt and backplate assembly by following these procedures:

WARNING

The backplate is under extreme pressure when the bolt is in the rear position. To avoid serious injury, ensure the bolt is forward before removing the backplate pin.

(a) Place the safety switch in the F (FIRE) position.

(b) Take out the backplate pin using the rim of a spent cartridge case or metal link.

(c) Pry outward on the pin lip and remove the pin with the fingers (Figure 2-18). If the pin cannot be removed with the fingers, hit the small end of the pin with the secondary drive lever until the pin comes loose.

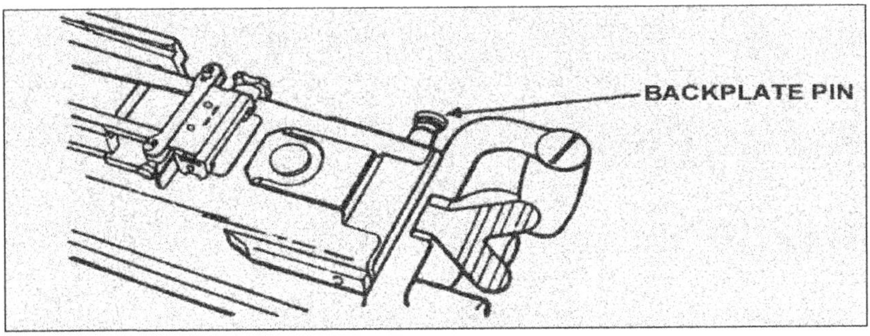

Figure 2-18. Backplate pin.

(d) Grasp the control grips with both hands and lift up slightly to disengage the backplate from the locking lugs in the receiver.

(e) Pull the bolt and backplate assembly to the rear.

(f) Once the bolt clears the sear, catch the bolt in one hand to prevent damage to the backplate assembly (Figure 2-19).

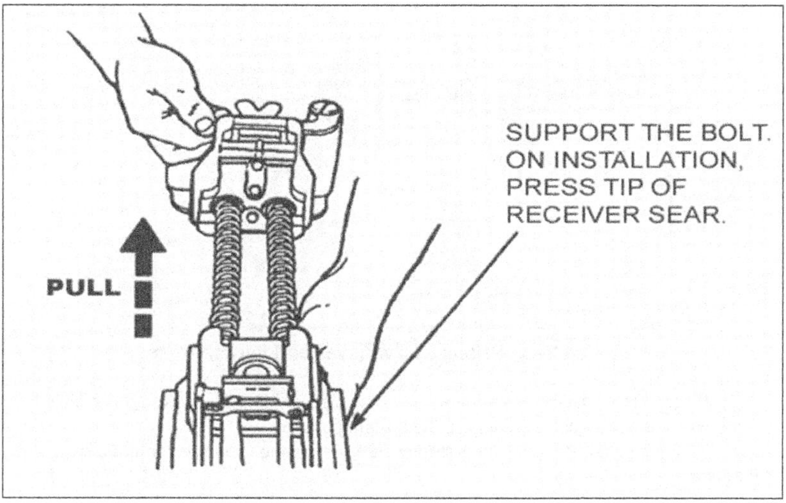

Figure 2-19. Removal of bolt and backplate assembly.

(9) Take off the primary drive lever and vertical cam.

(a) Grasp the primary drive lever and vertical cam with one hand and ensure they do not fall when released.

(b) Reach under the top of the receiver and locate the drive lever lock.

(c) Slide the lock one-quarter of an inch to the rear.

(d) Press down on the primary drive lever pivot post, which releases both the primary drive and vertical cam.

(e) Pull the primary drive lever from the front of the weapon and the vertical cam from the back (Figure 2-20).

CAUTION

Ensure you do not scratch the shiny portion of the vertical cam and primary drive lever. These areas have been specially treated to reduce friction; any scratches or use of abrasives will take off the treatment and increase wear.

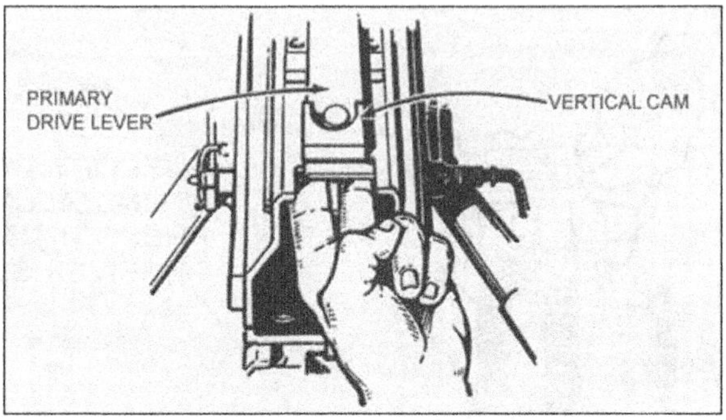

Figure 2-20. Primary drive lever and vertical cam.

(10) Take out the charger assemblies from both sides by following these procedures.

(a) Place the charger assemblies in the upright position.

(b) Using a metal link or spent cartridge case, retract the lock plunger at the base of the charger arm.

(c) Slide the charger housing rearward to disengage the lugs from the keyslots in the receiver.

(d) Lift the charger assembly away from the receiver (Figure 2-21).

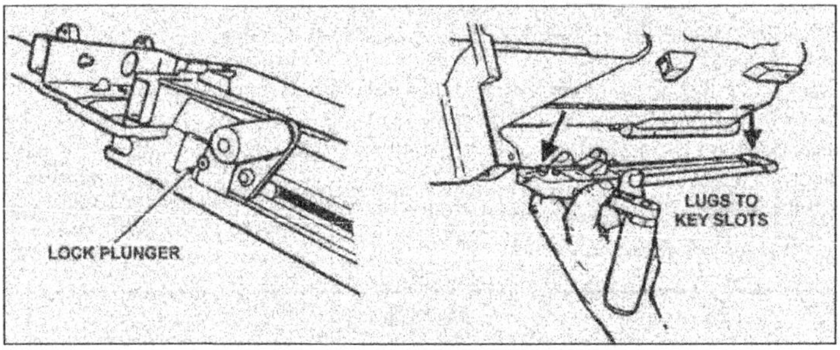

Figure 2-21. Charger assemblies.

(11) Take off the sear assembly by following these procedures:

(a) Turn the MK 19 on its side or upside down.

(b) Move the safety selector switch to F (FIRE), and use the rim of a spent cartridge case to lift up the sear lock plunger.

(c) At the same time, squeeze the sear and rotate the sear assembly (left or right) until you can no longer hold the sear (Figure 2-22).

(d) Release the plunger, and apply downward pressure to the sear assembly. (A spring in the sear will push the sear away from the receiver preventing further turning of the sear.)

(e) With the sear pressed flush to the receiver, continue to turn the sear until you hear a metallic click.

(f) Place the safety selector switch on S (SAFE) and continue to rotate the sear until it is at a 90-degree angle to the receiver.

(g) Lift the sear straight off the receiver.

(h) Place the safety switch in the S (SAFE) position.

Figure 2-22. Sear assembly.

b. **Assembly.** To assemble the gun, replace the groups in the exact reverse order in which they were taken off. Be sure the components are lubricated prior to reassembly.

(1) Install left and right hand charger assemblies by following these procedures:

(a) Turn receiver upright.

(b) Rotate charger handle to the forward straight-up position.

(c) Line up lugs on charger with slots in receiver rail. Insert charger lugs into slots.

(d) Hold charger tightly against rail. Slide charger forward until it locks in place.

(2) Install round-positioning block. Insert pins into the slots on the receiver and slide the block forward until the pins lock into the small sections of the holes.

(3) Install the ogive plunger assembly. Insert ogive plunger as shown in Figure 2-16.

(4) Install alignment guide assembly by following these procedures:

(a) Position the alignment guide assembly so that the pin is lined up with the slot in the feeder wall.

(b) Hold the alignment guide against the front wall and slide the alignment guide into the receiver until it locks or 'clicks.'

(5) Install feed tray and feed slide assembly by following these procedures (Figure 2-23):

(a) Place tray into top of feeder, recessed side up. The tabs on the feed slide should align with the slots on the feed slide tray.

(b) The hinge pinholes on the tray should line up with the hinge pinholes on the receiver.

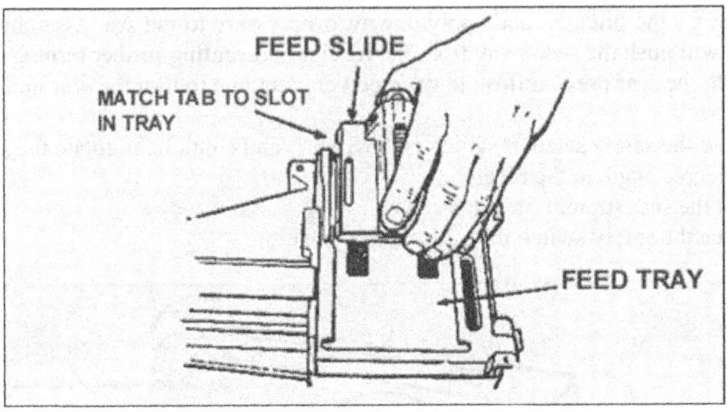

Figure 2-23. Feed slide assembly.

(6) Install top cover assembly by following these procedures:

> **CAUTION**
> To avoid breaking the crosspin, be sure it is fully inserted into the receiver before closing the top cover.

(a) Feed tray should be in place resting on the receiver.

(b) Place the top cover on the receiver with the hinge pinholes in line with the receiver lug end feed tray hinge pinholes.

(c) Hold top cover at a right angle to the receiver so that the key slots are aligned. Insert top cover pins on both sides. Ensure the crosspin is fully inserted then rotate the top cover fully open.

(7) Install secondary drive lever by following these procedures (Figure 2-24):

> **CAUTION**
> If the secondary drive lever is not properly engaged with the feedslide pin, the gun will not fire.

(a) Rotate the feed slide assembly and tray upward.

(b) Engage forked end of secondary drive lever with the feed slide pin.

(c) Press raised pivot post through hole in top cover.

(d) Press secondary drive lever against top cover until it locks in place.

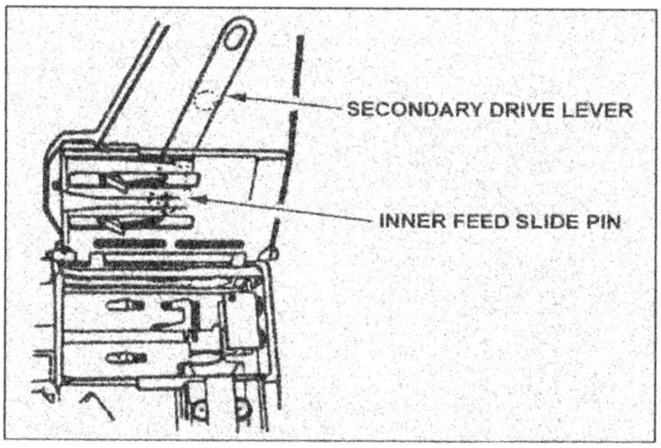

Figure 2-24. Installation of the secondary drive lever.

(8) Install vertical cam assembly by following these procedures:

> **CAUTION**
> Do not damage vertical cam by hitting it on the inside of the receiver.

(a) Slide vertical cam assembly through rear of receiver. Raised portion should slide over the top of the receiver. Drive lever lock should be underneath.

(b) Engage forked end in the notch.

(9) Engage primary drive lever by following these procedures:

(a) Hold vertical cam assembly in place and slide primary drive lever into receiver and rest the front end on the feed area.

(b) Slide drive lever lock rearward and engage pivot post of lever through holes in receiver and vertical cam.

(c) Slide drive lever lock (on the vertical cam just beneath top of receiver) forward.

(10) Install bolt and backplate assembly. With the sear assembly on the gun, assemble bolt and backplate using the following procedure (Figure 2-25):

> **CAUTION**
> Before inserting assembly, put cocking lever in forward position.

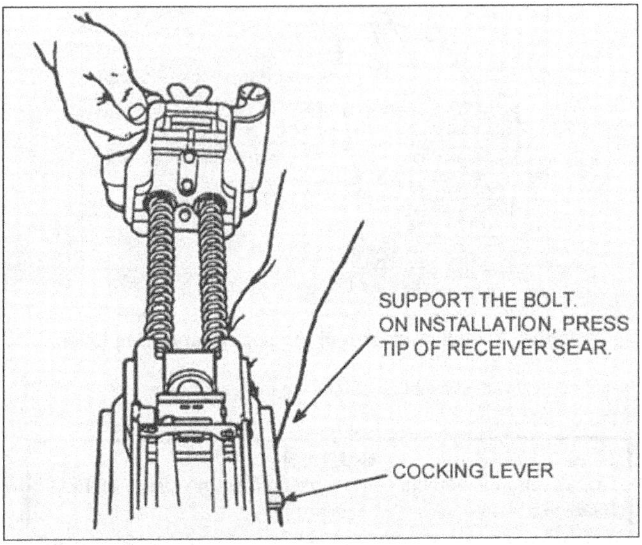

Figure 2-25. Installation of the bolt and backplate assembly.

(a) Place safety in F (FIRE) position.

(b) Ensure that the cocking lever is forward and align the bolt at the rear of the receiver and slide the receiver forward until it stops.

(c) Press down on the bolt release, slide the bolt slightly forward, and remove hands from the bolt release.

(d) Continue to slide the bolt forward until it stops again.

(e) Again, press down on the bolt release, slide the bolt slightly forward, and remove hands from the bolt release.

(f) Continue to slide the bolt forward until it reaches its most forward position.

(g) Seat the bottom of the backplate into the slots on the bottom of the receiver.

(h) Insert backplate pin to lock assembly in place.

(11) Install sear assembly by following these procedures:

(a) Turn receiver over on its top.

(b) Place sear housing on the receiver and line up sear housing assembly at a right angle to the barrel centerline.

(c) Put safety on F (FIRE) position.

(d) Press down and rotate the sear assembly until it locks in place.

(12) Install feed throat assembly by following these procedures:

(a) Squeeze the plungers, and align the pins with the holes in the receiver.

(b) Release plunger to reattach feed throat.

NOTE: Ensure the safety switch is in the F (FIRE) position so the sear can be easily depressed.

2-5. LOADING PROCEDURES

Before loading, the gunner should ensure the MK 19 is on S (SAFE) and the bolt is in the forward position.

a. Before loading, use the following procedures:

(1) Attach the feed throat by squeezing the spring-loaded pins on the feed throat (Figure 2-26).

(2) Insert the feed throat on both sides of the feeder.

(3) Ensure that the feed throat points down.

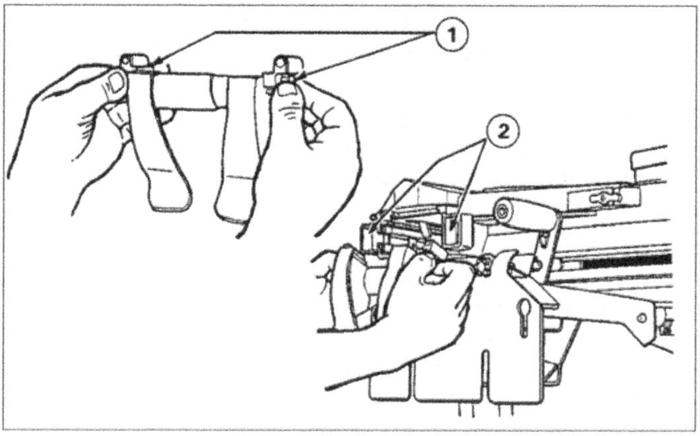

Figure 2-26. Attachment of feed throat.

b. When loading, use the following procedures:

(1) Insert the first round into the feeder (female link first) (Figure 2-27).

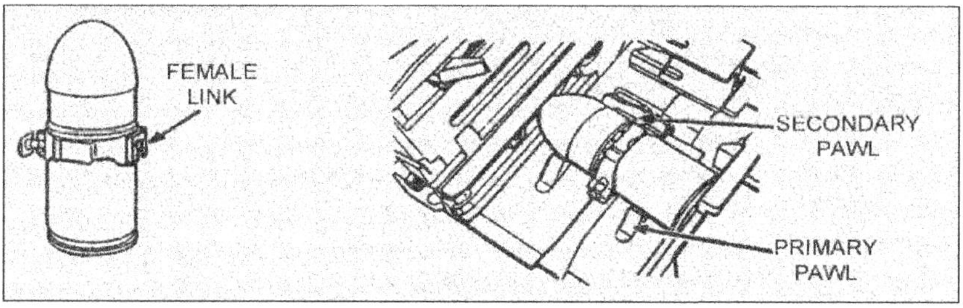

Figure 2-27. Loading of first round.

(2) Push the round across the secondary feed pawl. To move the feed slide to the left, push the secondary drive lever to the right (Figure 2-28).

(3) Close the cover.

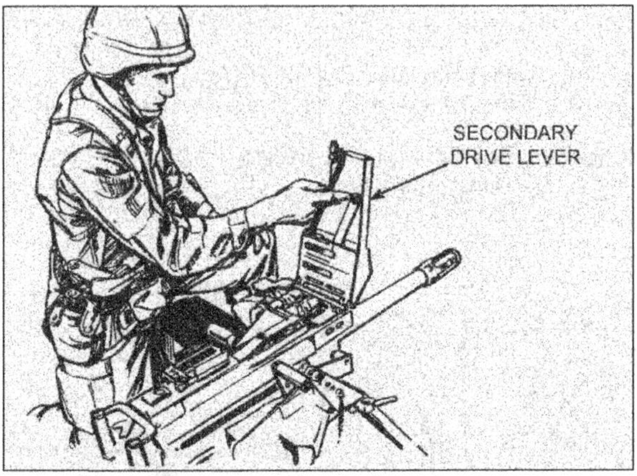

Figure 2-28. Positioning of feed slide assembly.

(4) Grasp the charger handles with the palms down.

(5) Press the charger handle locks in.

(6) Rotate the handles down and pull them sharply to the rear.

(7) Return the charger handles forward to their original upright position, after locking the bolt to the rear.

CAUTION

Failure to completely pull the bolt to the rear may result in the misalignment of the M16A2 links on the round, which causes the round to feed improperly.

(8) Place the safety switch on F (FIRE) and press the trigger. The bolt slams forward and grasps the first round in the bolt extractors.

(9) Grasp, unlock, and turn charger handles down, and lock the bolt to the rear again.

(10) Ensure the safety switch is on S (SAFE).

(11) Return the charger handles to their original upright position. The MK 19 is now ready to fire.

2-6. MALFUNCTIONS, STOPPAGES, AND CORRECTIONS

A malfunction is a failure of the weapon to function properly. Neither defective ammunition nor improper operation of the gun by a crewmember is considered a malfunction of the MK 19. A stoppage is any interruption in the cycle of operation caused by faulty action of the weapon or ammunition. Appendix C, Stoppages, Malfunctions, Immediate Actions, and Remedial Actions, provides inspections, corrective actions, and emergency actions of malfunctions and stoppages. For more information on troubleshooting common malfunctions and stoppages, refer to TM 9-1010-230-10.

2-7. CARE AND CLEANING

The MK 19 requires preventive maintenance checks and services (PMCS) and lubrication before, during, and after firing. It should be checked and cleaned daily when used and not fired, and weekly if not used. For more detailed guidance on the care, cleaning, and PMCS for the MK 19, refer to TM 9-1010-230-10.

a. **Cleaning and Lubrication.** Several types of lubricants can be used on the MK19. High friction areas of the gun have a protective coating and care must be taken to prevent its removal. Listed below are authorized lubricants, and other information about lubricating and cleaning the MK19.

(1) ***Authorized Lubricants.*** These are the authorized lubricants used on the MK 19:

(a) Lubricant, artic weather, automatic weapons (LAW).

(b) Lubricating oil, semifluid, automatic weapons (LSA).

(c) Lubricating oil, semifluid, automatic weapons, temperate (LSAT).

(d) Rifle bore cleaner (RBC). RBC should be used sparingly only in the bore and when necessary.

(e) Grease, molybdenum disulfide (GMD).

(2) ***Use of Lubrications.*** The following is general guidance on the use of lubricants on the MK 19.

(a) Never mix lubricants on the MK19. If you change lubricants or do not know for sure which lubricant is currently applied to the MK19, completely clean off all old materials before you apply a new lubricant.

(b) Never use cleaner lubricant petroleum (CLP) on a MK19. It gums up moving parts instead of lubricating them.

(c) Apply a heavy coat of lubricant and do not wipe it off.

(d) In cold weather, 0 degrees F to -25 degrees F, use LSAT, GMD, or LAW.

(e) In extra cold weather, -25 degrees F and below, use LAW.

(3) ***High Friction Areas.*** The under side of the primary drive lever, the vertical cam assembly, the bolt (where it rides on the rails), and the area where the feed slide and feed slide tray connect have been treated to prolong the life of the MK 19. Special care is required during cleaning and maintenance to prevent damage to the coating. CLP, acids, and abrasives should never be used on the MK 19 because they will remove the protective coating.

b. **General Cleaning and Lubrication Instructions.** The following is general guidance for the cleaning of the MK 19.

(1) Wipe or brush off dirt and grime.

(2) When lubricating the weapon, give extra attention to the feed pawls, the cocking lever rails, the pivot posts on the primary drive lever, and the bolt assembly.

(3) Lubricate each part. Give special care to those hard-to-get spots.

(4) Work in the lubricant by moving the parts.

c. **When to Clean, Lubricate and Inspect the MK 19.** Clean and inspect the gun thoroughly. Report all worn, burred, defective, or missing parts to your armorer or support maintenance. In general, lubricate and clean the MK19 as follows:

(1) Always clean and lubricate the gun after firing.

(2) Clean and lubricate the gun daily if it is used but not fired.

(3) Clean and lubricate the gun weekly if the weapon is stored.

d. **Inspections.** Several parts of the MK 19 should be checked whenever possible. These include:

(1) The vertical cam. Check closely for smoothness.

(2) The gap between the charger assemblies and the receiver.

(3) All parts for wear, cracks, missing items, and abnormal shapes.

2-8. OPERATION UNDER ADVERSE CONDITIONS

The MK 19 is a robust weapon and designed to operate over a wide range of weather conditions but additional care is required when the temperatures are hot or cold, and the conditions are wet, sandy, or icy.

a. During *hot, wet, or salty air conditions* follow these procedures:

(1) Inspect the weapon more frequently for signs of rust.

(2) Keep gun as moisture-free as possible.

(3) Field strip, clean, and lubricate more often to preserve metal and prevent rust.

(4) Use a generous second coat of lubrication for extra protection.

b. During *hot, dry, sandy, or dusty conditions* follow these procedures:

(1) Do not lubricate the entire exposed metal surfaces as this will only collect dust and sand.

(2) Clean the weapon thoroughly and lubricate only the moving components.

(3) Extreme heat dries up lubricant. Clean and lightly lubricate the moving components of the weapon more frequently.

c. During *cold, icy, or snowy conditions* follow these procedures:

(1) Prior to using cold-weather lubricant, be sure to clean off all existing lubrication. From 0 degrees F to –25 degrees F, use either LSAT, GMD, or LAW. At temperatures –25 degrees F and below, use LAW.

(2) If the weapon 'sweats,' dry and lube the parts before taking the weapon outdoors and apply a light second coat to provide protection.

(3) Cover weapon if left outside.

(4) If possible, thoroughly clean, dry, and lubricate the weapon in a warm place.

(5) If the weapon is brought indoors, keep it away from direct heat.

(6) Perform functional checks and lubricate daily to help prevent corrosion.

2-9. DESTRUCTION

Only the commander can direct the destruction of the MK 19. There are several ways to destroy it. Methods include destruction by mechanical means, burning, gunfire, demolition or disposal. Use the same methods to destroy equipment. However, ammunition must be handled differently. The best way to destroy ammunition is to fire it. Otherwise, dispose of it by burial, dumping it in a body of water, or using the demolition method.

WARNING

Ammunition must never be scattered (this may result in a hazard to friendly soldiers), burned (HE and HEDP rounds may explode), or destroyed mechanically.

a. **Mechanical Method.** If possible, use the mechanical method to destroy MK 19s and equipment. There are three ways to do this:

(1) Smash them. Using a sledgehammer or some similar heavy tool, damage the MK 19s so that they cannot be used.

(2) Bend the MK 19s' guide rods over the receivers.

(3) Take off (or out) the same part from each of the MK 19s. If a different part is taken from each gun, new MK 19s may be pieced together.

b. **Burning Method.** To burn equipment or MK 19s, use a fuel source that burns hot enough to take the temper from the metal. For example, use jet propulsion fuel (JP-4), oil, or gas. All friendly soldiers in the area are warned before equipment or MK 19s are bare burned, because a fire will reveal the positions of friendly forces.

c. **Gunfire Method.** This is the least reliable way to destroy MK 19s or equipment. Pile up all of the equipment and MK 19s and use MK 19s, grenades, or rifles to fire on them. They will scatter so check after shooting to see that they are all destroyed.

d. **Demolition Method.** Use an explosive charge such as composition 4 (C4) or any type of HE round. Place the charges so they will completely destroy all the MK 19s and equipment.

e. **Disposal Method.** To dispose of MK 19s and equipment, break them down and either scatter the parts into swamps or marshes, or bury them.

(2) Push the round across the secondary feed pawl. To move the feed slide to the left, push the secondary drive lever to the right (Figure 2-28).

(3) Close the cover.

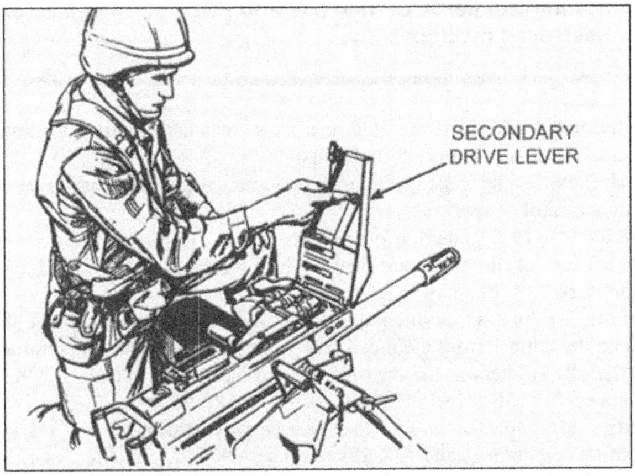

Figure 2-28. Positioning of feed slide assembly.

(4) Grasp the charger handles with the palms down.

(5) Press the charger handle locks in.

(6) Rotate the handles down and pull them sharply to the rear.

(7) Return the charger handles forward to their original upright position, after locking the bolt to the rear.

CAUTION

Failure to completely pull the bolt to the rear may result in the misalignment of the M16A2 links on the round, which causes the round to feed improperly.

(8) Place the safety switch on F (FIRE) and press the trigger. The bolt slams forward and grasps the first round in the bolt extractors.

(9) Grasp, unlock, and turn charger handles down, and lock the bolt to the rear again.

(10) Ensure the safety switch is on S (SAFE).

(11) Return the charger handles to their original upright position. The MK 19 is now ready to fire.

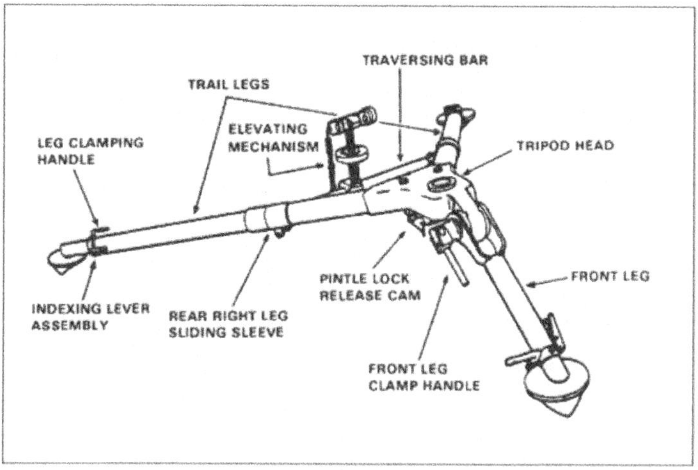

Figure 3-2. M3 tripod.

(1) Unscrew the leg-clamping handle; press down on the indexing lever, and extend the leg to the desired length.

(2) Align the indexing lever stud with one of the holes in the tripod leg extension.

(3) Release pressure on the indexing lever, allowing the stud to fit the desired hole. Tighten the leg-clamping handle.

b. Set the front leg of the tripod by following these procedures:

(1) Turn the front leg clamp handle counterclockwise to loosen the front leg.

(2) Adjust the leg to the desired angle, ensuring the tripod head is level, and tighten the front leg clamp.

c. Secure the tripod legs by following these procedures:

(1) Stamp the metal shoe on each tripod leg into the ground.

(2) Sandbag each leg to stabilize the MK 19 for firing.

d. Mount the M64 gun cradle onto the M3 tripod by following these procedures:

(1) Unlock the tripod pintle lock release cam.

(2) Insert the gun cradle's pintle into the tripod pintle bushing (Figure 3-3).

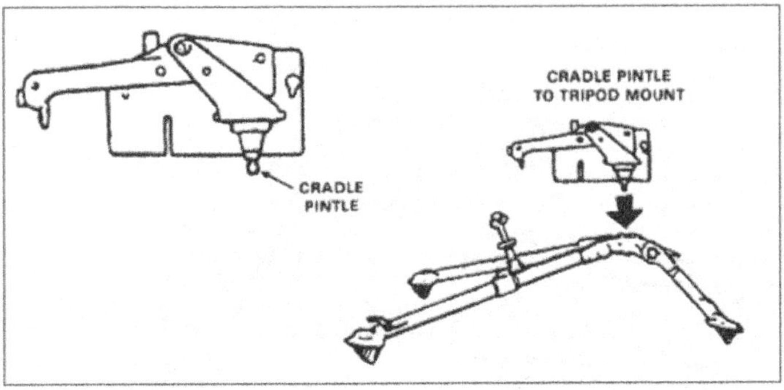

Figure 3-3. Insertion of pintle into pintle bushing.

(3) Lock the pintle lock release cam to secure the gun cradle. Check the gun cradle, by pulling up on it slightly, to ensure that it is seated and locked.

e. Attach the (traverse and elevating) T&E mechanism by following these procedures (Figure 3-4):

(1) Zero the T&E mechanism by zeroing the traversing handwheel, elevating handwheel, and the elevating mechanism sleeve to the lower elevating screw.

(a) To zero the traversing handwheel, hold the T&E mechanism so that the traversing handwheel is on the left when looking at it, then turn the traversing handwheel toward you until it stops. Loosen the locking nut slightly. Align the zero on the scale with the zero on the elevating screw yoke. Hold the scale with the zeros aligned, and tighten the locking nut. Make sure the zeros stay aligned. Turn the traversing handwheel two complete revolutions away from you. If doing this at night, count 50 "clicks" away from you.

(b) To zero the elevating handwheel to the upper elevating screw, align the two zeros. Rotate the elevating handwheel up or down until a zero with a line below it is visible on the upper elevating screw. Position the elevating handwheel so the indicator is pointing at the zero on the handwheel.

(c) Zero the elevating mechanism sleeve to the lower elevating screw. Rotate the elevating mechanism sleeve all the way up; rotate it down until it stops and note the number of complete turns. Rotate the elevating mechanism sleeve up half that number of turns. Position the slide lock lever to face you.

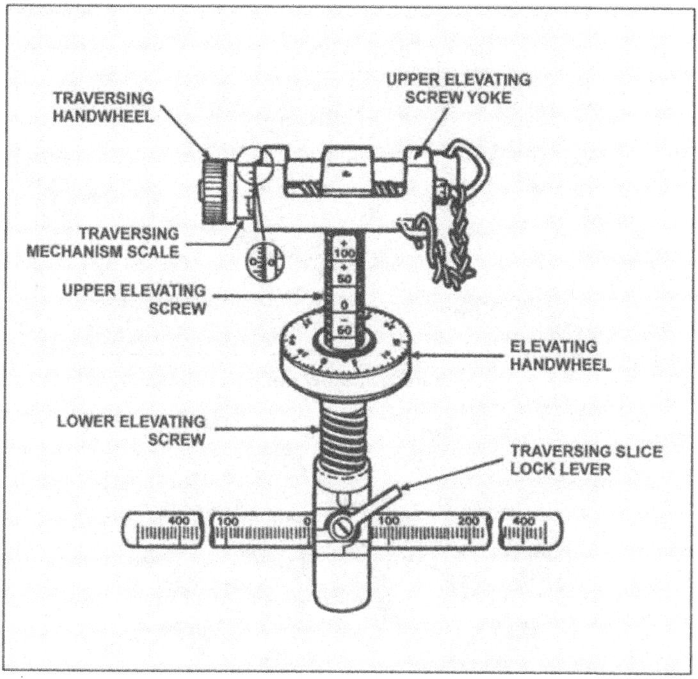

Figure 3-4. T&E mechanism.

(2) Remove the stow pin from the gun cradle (Figure 3-5).

(3) Align holes in the upper elevating screw yoke of the T&E mechanism with the rear holes in the gun cradle (Figure 3-6).

NOTE: The stow pin locks the cradle in a horizontal position, preventing it from depressing or elevating.

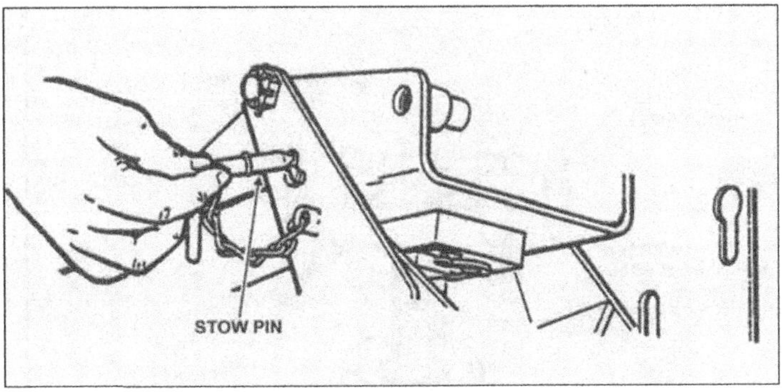

Figure 3-5. Stow pin.

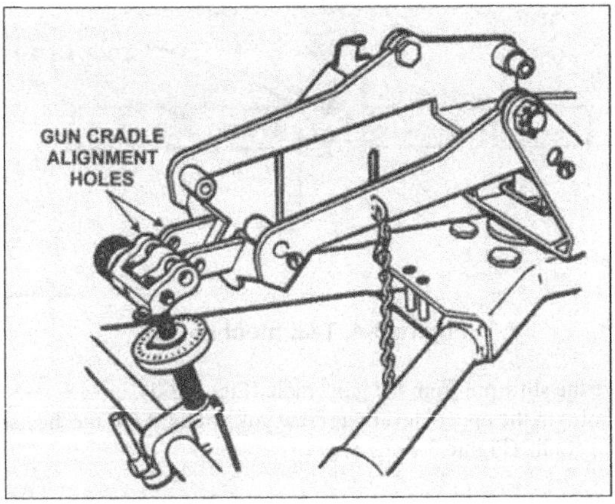

Figure 3-6. Alignment of gun cradle to T&E mechanism.

(4) Lock the elevating sleeve mechanism onto the center of the traversing bar.

(5) Insert the quick-release pin from the right (Figure 3-7).

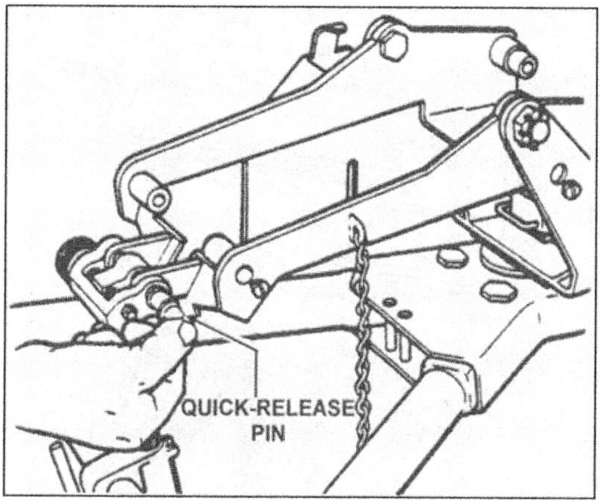

Figure 3-7. Quick-release pin.

f. Mount the MK 19 by following these procedures:

(1) Lift the MK 19 into the gun cradle. One soldier holds the barrel while another holds the control grips. The one holding the grips stands between the trail legs of the tripod; the other straddles the front leg. The soldiers lift the MK 19, barrel first, onto the cradle.

(2) Align the grooves on the receiver with the lugs in the gun cradle, and slide the receiver forward (Figure 3-8).

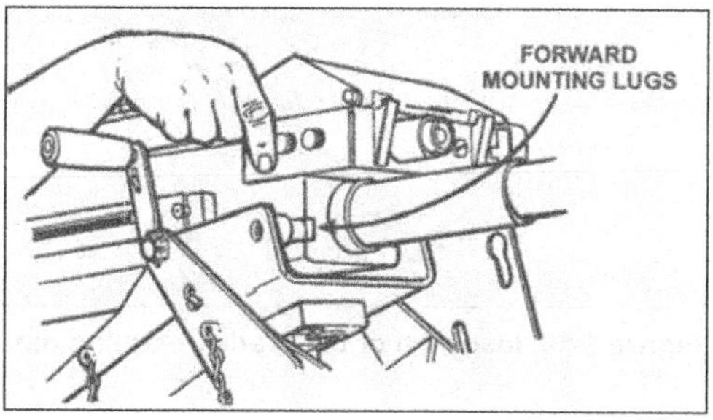

Figure 3-8. Mounting lugs.

(3) Align the sear mounting holes with the gun cradle mounting holes (Figure 3-9). Secure the rear of the receiver by inserting the retaining pin through the cradle and sear assembly and rotate it until it locks in place (Figure 3-10). If a safety clip is attached, use it to secure the retaining pin.

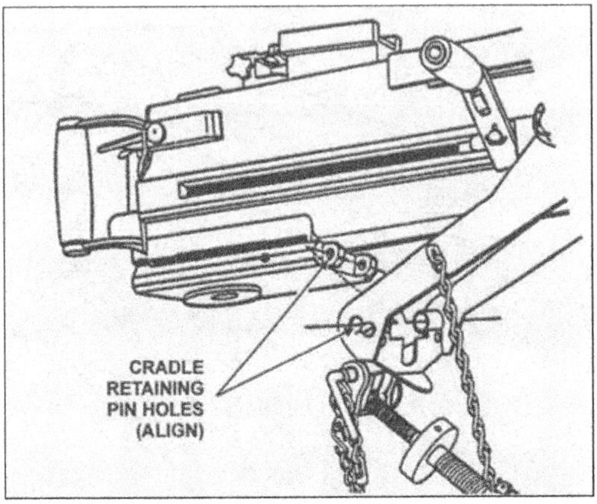

Figure 3-9. Alignment of the sear assembly and pinholes.

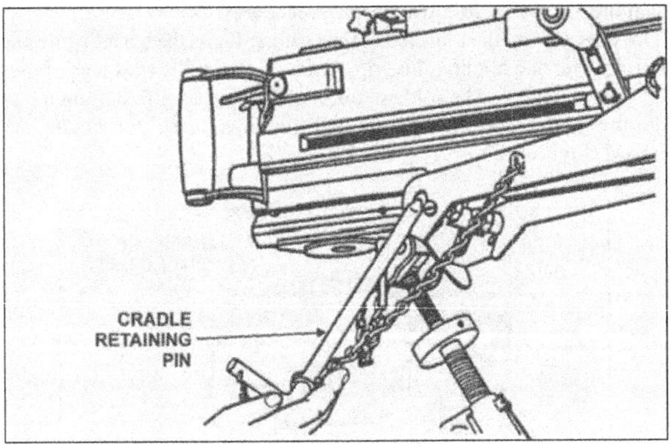

Figure 3-10. Insertion of the cradle retaining pin.

g. Attach the feed throat to the MK 19 by following these procedures:

(1) Squeeze together each set of grip pins (Figure 3-11).

(2) Attach the feed throat to the front left-hand side of the receiver assembly. The pins of the feed throat must line up with the pinholes in the receiver (Figure 3-12).

(3) Relax pressure on the spring-loaded grip pins so they will snap into place (Figure 3-13, page 3-9).

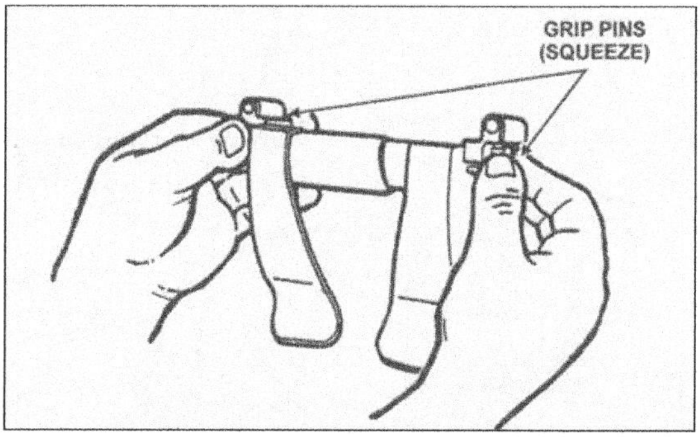

Figure 3-11. Grip pins.

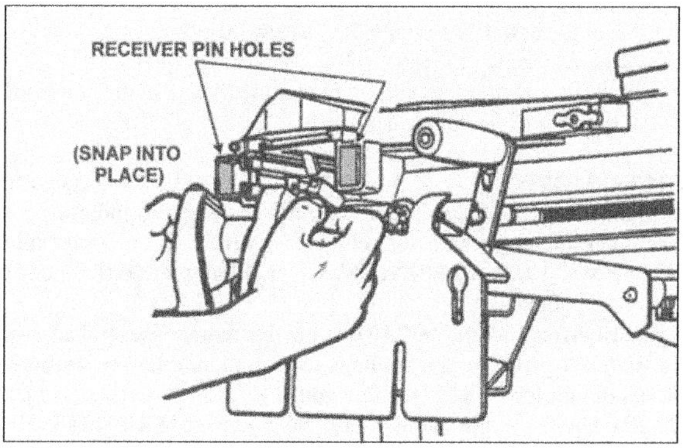

Figure 3-12. Feed throat alignment.

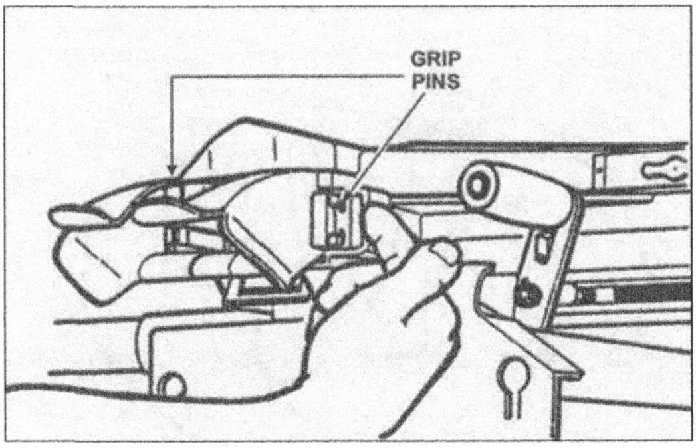

Figure 3-13. Feed throat installation.

NOTE: 1. When the MK 19 is mounted on the M3 tripod or any vehicle, it is called a weapon system.
2. The same procedure is used to mount the MK 19 to the gun cradle regardless of the type of mount used.

3-3. VEHICLE-MOUNT

This section lists the accessories used and outlines the steps to mount the MK 19 onto vehicles. Since mounting procedures for vehicles vary little, this section explains in detail how to mount the MK 19 on the HMMWV and highlights the differences in mounting procedures for other vehicles.

a. **Accessories.** Mounting the MK 19 on vehicles requires several accessories.

(1) ***Gun Cradle.*** The MK 64 gun cradle is used to mount the MK 19 onto any vehicle having a pedestal mount for the M2 machine gun.

(2) ***Pintle Adapter.*** The pintle adapter is used to mount the MK 19 in all vehicular modes (Figure 3-14). Other accessories are used to mount the MK 19 to specific vehicles. The upper end accepts the gun cradle's pintle, which is secured by a quick-release pin. The lower end of the adapter fits the mounting wells on the M4 pedestal, HMMWV weapon platform, M36A2 ring mount with M66 ring, and commander's cupola on the M113 APC.

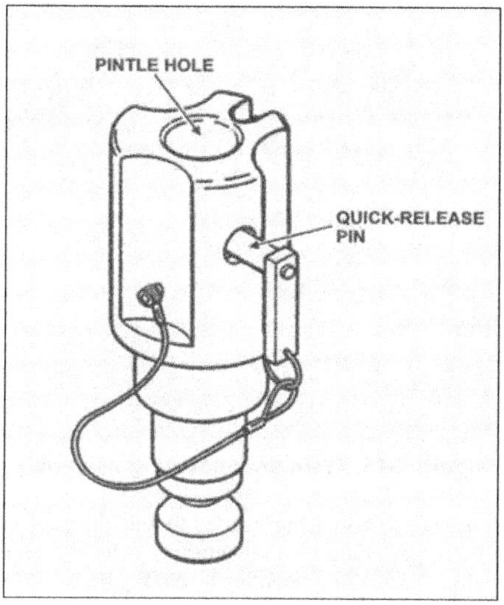

Figure 3-14. Pintle adapter.

(3) ***Train and Elevating Assembly.*** This assembly secures the MK 64 gun cradle to the M4 pedestal or HMMWV weapon platform pedestal and allows directional adjustments. The lower end of the train and elevating assembly is attached to the pedestal by a train lock clamp. The clamp is released or locked in position by a train lock handle. Two positioning clamps are supplied to prevent vertical movement of the train lock clamp on the M4 pedestal. When used on the HMMWV weapon platform pedestal, *only one clamp is needed above the train lock clamp.* The upper end of the train and elevating assembly is a standard caliber .50 T&E mechanism and it is attached to the lower rear holes in the M4 cradle by a retaining pin (Figure 3-15).

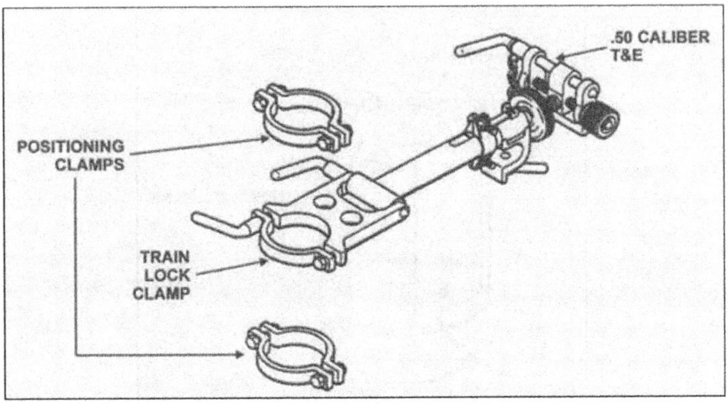

Figure 3-15. Train and elevating assembly.

(4) ***Travel Lock Adapter.*** This adapter is used to hold the gun cradle securely during travel. It is attached to the gun cradle by a retaining pin, which is inserted through the gun cradle train and elevating mechanism mounting holes (Figure 3-16).

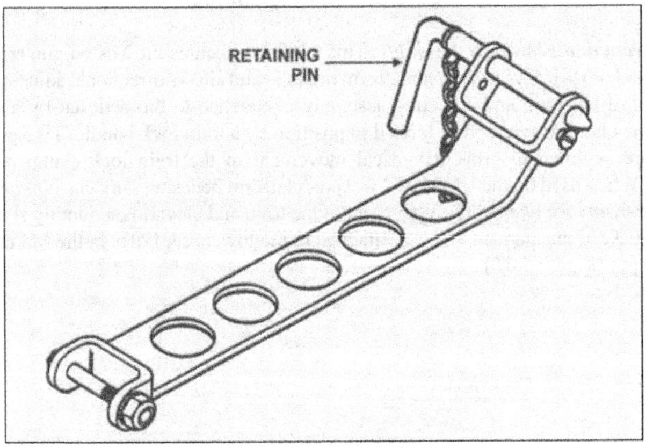

Figure 3-16. Travel lock adapter.

(5) ***Travel Lock Bracket.*** This bracket allows the travel lock adapter to be used with the M4 pedestal mount. It has two halves that are secured around the pedestal with nuts and bolts. The bracket is attached to the travel lock adapter by aligning the lower holes (Figure 3-17).

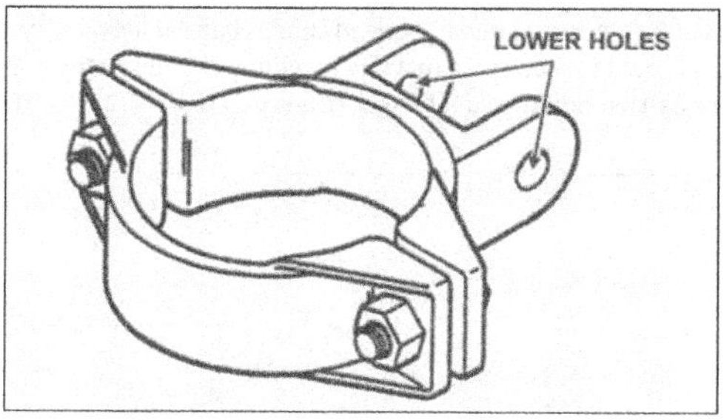

Figure 3-17. Travel lock bracket.

(6) ***Bracket Mounting Assembly.*** This assembly supplies a mount for the M548 metal ammunition container. It has a metal frame that attaches to the gun cradle, and a retaining pin that inserts through the top-inner end of the M548 ammunition container (Figure 3-18).

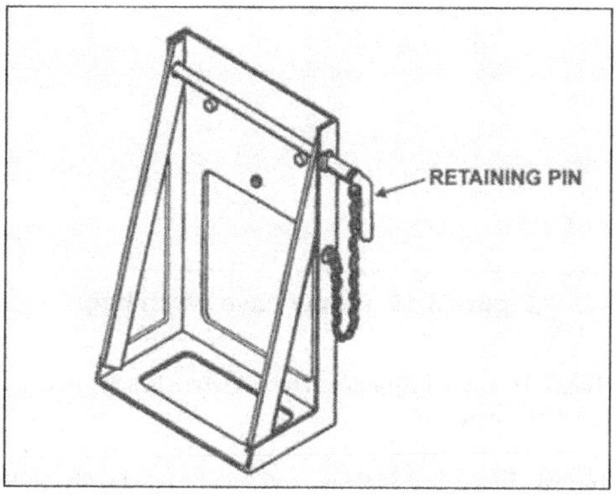

Figure 3-18. Bracket mounting assembly.

(7) ***Empty Case Catch Bag.*** This rubberized canvas bag catches empty cartridge cases as they are ejected. It is held in place by a metal rim, which attaches to the bottom of the cradle under the receiver by two hooks and a hanger (Figure 3-19).

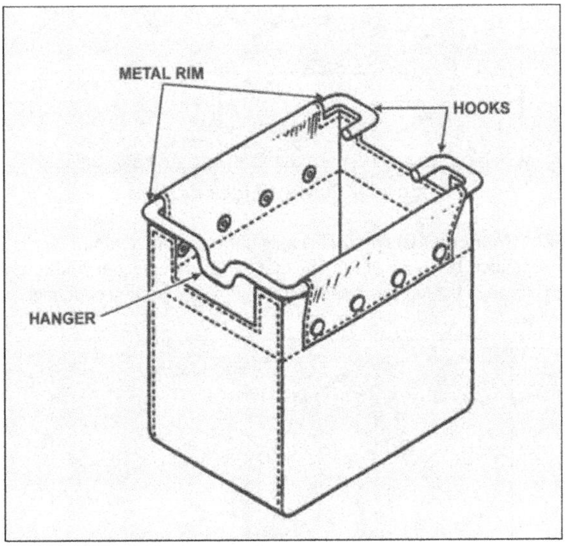

Figure 3-19. Empty case catch bag.

NOTE: The 9/16-, 7/16-, and 3/8-inch open-end wrenches are necessary tools to mount the MK 19 on vehicles.

b. **Vehicle Mounts.** Various vehicle mounts can be used with the MK 19.

(1) ***HMMWV Weapon Platform.*** To mount the MK 19 on the HMMWV weapon platform (Figure 3-20):

(a) Insert the pintle adapter onto the HMMWV pedestal by following these procedures:

(1) Loosen the HMMWV pedestal lock screws by turning counterclockwise until the threaded ends are flushed with the pedestal's inner wall (Figure 3-20).

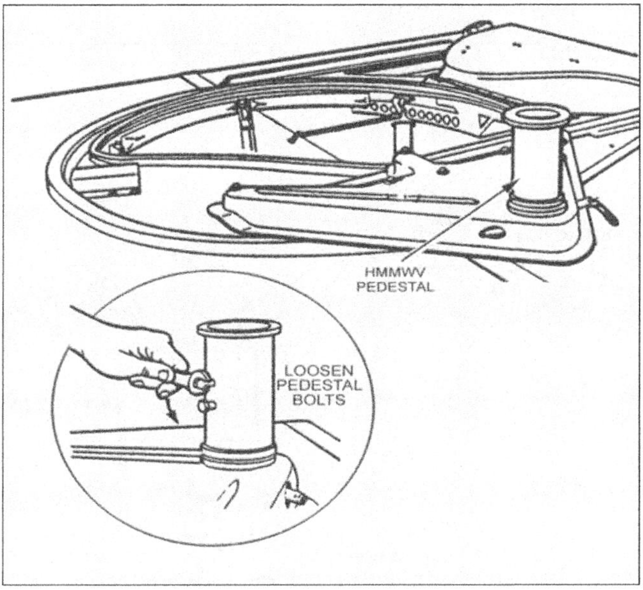

Figure 3-20. HMMWV pedestal.

(2) Insert the pintle adapter assembly by placing the lower end of the pintle adapter into the HMMWV pedestal (Figure 3-21).

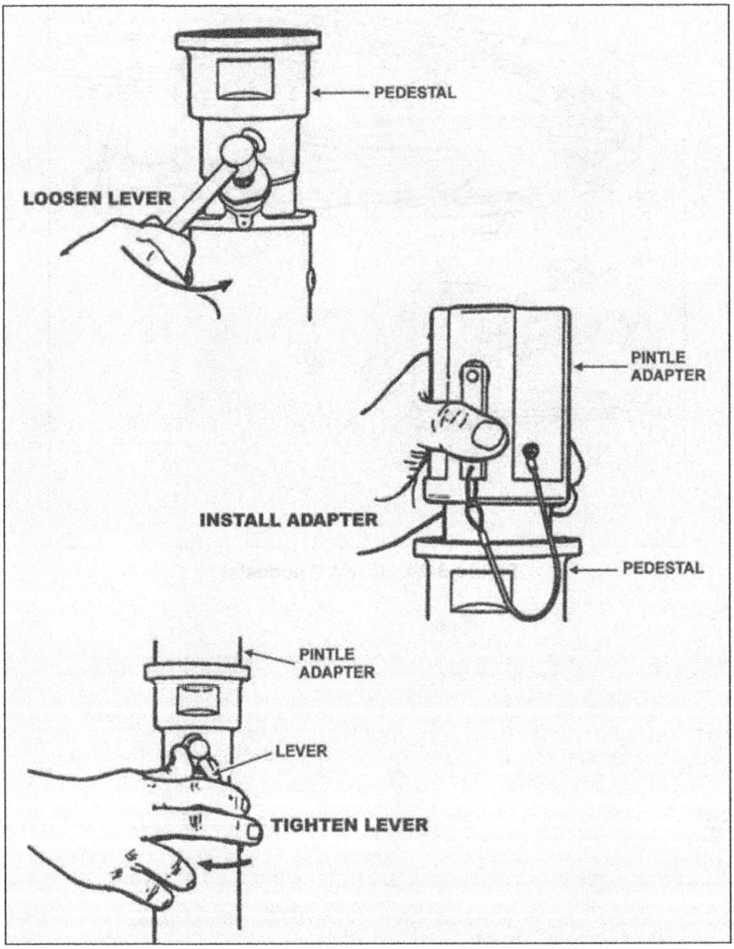

Figure 3-21. Pintle adapter assembly.

(3) Tighten the lock screws with a 3/8-inch open-end wrench. Pull upward on the pintle adapter to ensure it is secured in place.

(4) Remove the pintle adapter quick-release by pressing in on the pin's quick-release button and pulling the pin from the pintle adapter.

(b) Install the carriage and cradle assembly by following these procedures (Figure 3-22):

(1) Insert the front stow pin to prevent relative movement to the carriage and cradle.

(2) Insert the carriage pintle into the top of the pintle adapter assembly.

(3) Press in on the pin's quick-release button and insert the pin.

(4) Pull upward and twist the carriage and cradle assembly. It should lock into the pintle adapter, but it should traverse freely left and right.

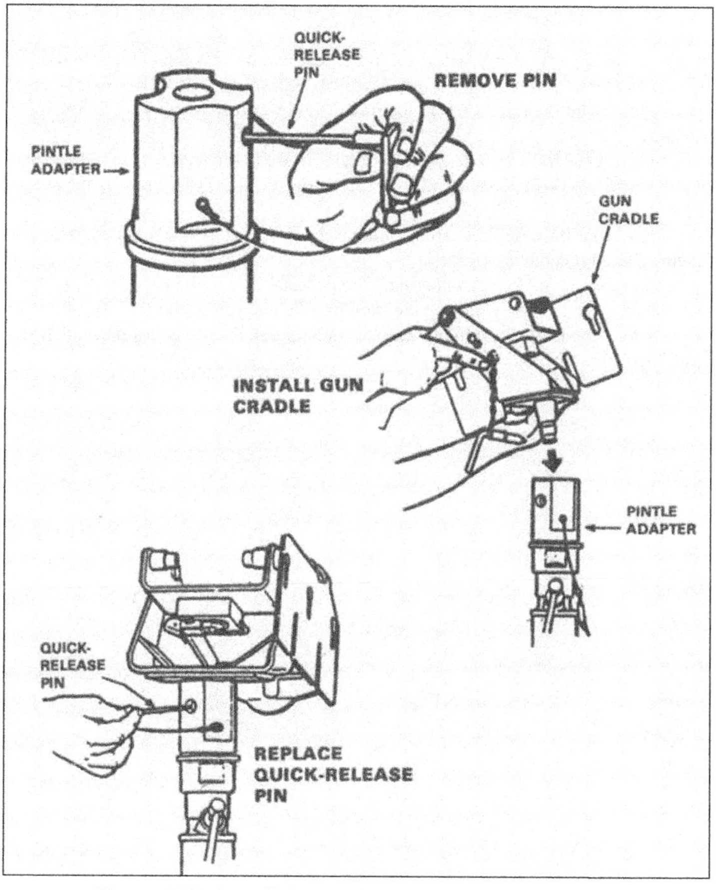

Figure 3-22. Install the carriage and cradle assembly.

(c) Install the T&E mechanism in the train and elevation assembly by following these procedures:

(1) Turn the elevating handwheel to read 250 mils.

(2) Loosen the pivot arm screws, using two 7/16-inch open-end wrenches.

(3) Insert the T&E mechanism all the way into the train and elevation assembly.

NOTE: The T&E mechanism is the upper portion of the training and elevation assembly; it must be installed in the train and elevation assembly *before* the training and elevation assembly can be installed on the vehicle.

(4) Position the traversing lock lever to the rear and the traversing knob to the left by tightening the pivot arm screws, using the two wrenches.

(d) Separate the middle clamp on the train and elevation assembly by following these procedures:

(1) Unscrew the train lock handle counterclockwise to remove.

(2) Using the 9/16-inch open-end wrench, remove the hex head screw on the other side of the middle clamp.

(e) Attach the train and elevation middle clamp to the HMMWV pedestal post by following these procedures (Figure 3-23):

(1) Assemble the middle clamp around the pedestal.

(2) Tighten the clamp by turning the train lock handle clockwise.

(3) Using the 9/16-inch open-end wrench, equally tighten the screw on the other side of the middle clamp.

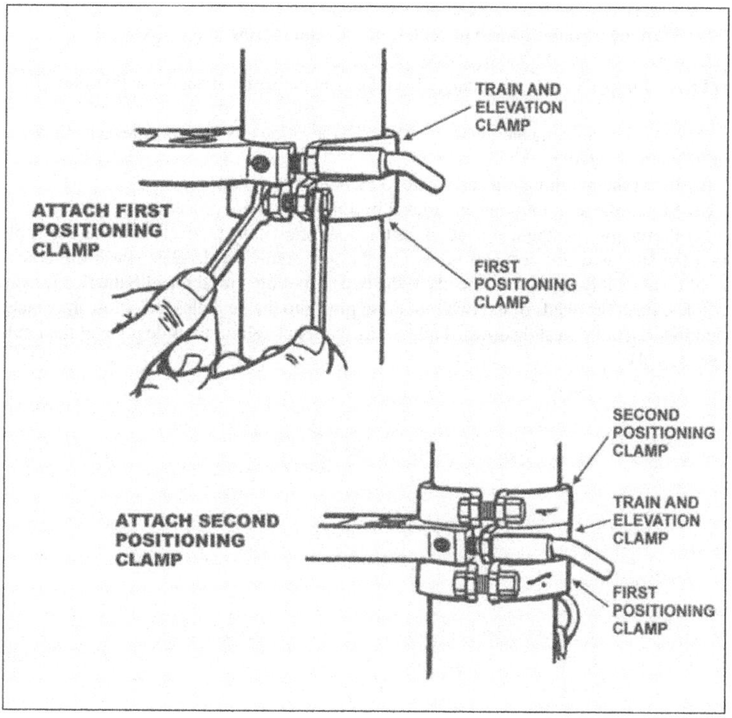

Figure 3-23. Train and elevation middle clamp.

(f) Install a support clamp above the middle clamp. To allow free left and right traverse of the carriage and cradle assembly, loosen the train lock handle. To lock in place, tighten the handle.

NOTE: The second or lower set of clamps is not used in this application since the train lock rests on the built-up base of the HMMWV pedestal. This will also allow for increased elevation. Using two 9/16-inch open-end wrenches, fasten one set of clamps above the middle clamp. Tighten each screw two turns until snug.

(g) Attach the train and elevation assembly to the carriage and cradle assembly by following these procedures:

(1) Pull out the front stow pin to allow the cradle to move.

(2) Pull out the train and elevation assembly's retaining pin.

(3) Ensure the elevating handwheel is set at 250 mils. Position the lock lever to the rear and ensure the traversing knob is on the left.

(4) Turn the traversing knob to center the elevating screw in the yoke.

NOTE: Ensure the elevating handwheel does not get jammed under the lip of the cradle.

(5) Turn the elevating handwheel to align the elevating mechanism holes with the lower rear holes in the cradle.

(6) Insert the retaining pin and rotate to the locked position.

(h) Mount the ammunition can bracket by following these procedures:

(1) Partly unscrew the wing nut on the threaded stud of the bracket mounting assembly. Align the stud with the forward groove in the gun cradle's side plate. Push the bracket mounting assembly up until the heads of the two mounting pins align with the two forward keyholes. Push the heads of the two mounting pins into the keyholes and allow the bracket mounting assembly to slide down. Tighten the wing nut behind the side plate of the cradle (Figure 3-24).

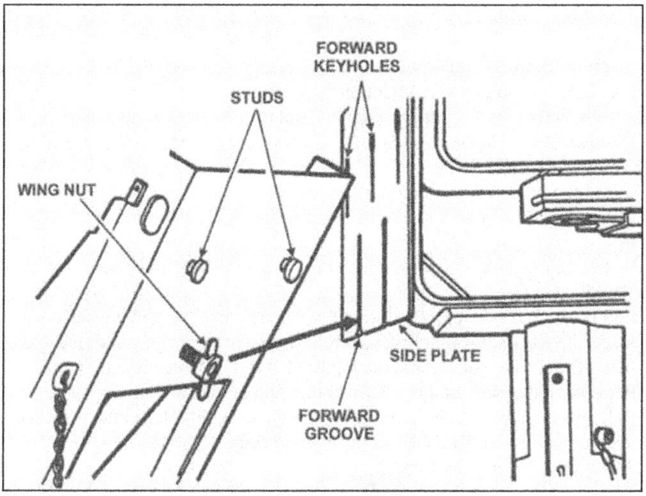

Figure 3-24. Attachment of bracket mounting assembly.

(2) Insert the two hooks on the empty case catch bag through the rear holes in the gun cradle. Engage the single front hanger on the catch bag with the hook on the gun cradle (Figure 3-25).

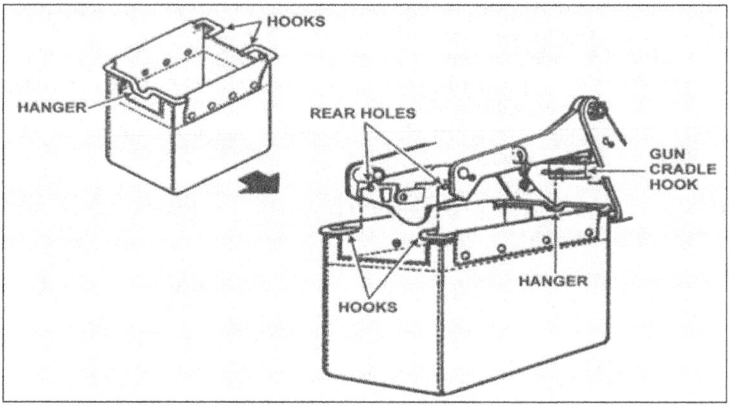

Figure 3-25. Attachment of empty case catch bag.

(2) ***M66 Ring Mount (2 1/2- to 5-ton cargo trucks).*** To mount the MK 19 on the M66 ring mount:

(a) Insert the pintle adapter into the M36A2 ring mount with the M66 ring receptacle.

(b) Install the gun cradle and mount the MK 19.

(c) Attach the bracket mounting assembly and the empty case catch bag.

(3) ***Commander's Cupola (M113 APC).*** To mount the MK 19 on the commander's cupola follow these procedures:

(a) Install the pintle adapter and the gun cradle.

(b) Mount the MK 19 and attach the bracket mounting assembly and the empty case catch bag.

(c) Install the machine gun pintle adapter in the commander's cupola mount by pushing the lock handle down.

(d) Pull the lock handle up, if it does not come up when the pintle is installed.

(e) Insert the front stow pin to prevent relative movement of the carriage and cradle.

(f) Insert the carriage pintle into the top of the pintle adapter assembly.

(g) Press in on the pin's quick-release button and insert the pin.

(h) Pull upward and twist the carriage and cradle assembly. It should be locked into the pintle adapter, but it should traverse freely left and right.

(i) To remove, reverse the steps.

(4) ***The M88 Recovery Vehicle Mount.*** To mount the MK 19 on the M88 recovery vehicle mount follow these procedures:

(a) Loosen the traverse locking screw.

(b) Install the machine gun pintle in the machine gun mount.

(c) Insert the front stow pin to prevent movement of the carriage and cradle. Insert the carriage pintle into the top of the pintle adapter assembly.

(d) Press in on the pin's quick-release button and insert the pin.

(e) Pull upward and twist the carriage and cradle assembly. It should be locked into the pintle adapter, but it should traverse freely left and right.

(f) To remove, reverse the steps.

(5) ***Weather Cover.*** To install the weather cover on the MK 19 follow these procedures:

(a) Unzip the zipper on the weather cover.

(b) Pull the weather cover over the barrel, ammunition can, and rear of the MK 19.

(c) Zip the zipper to secure.

NOTE: The weather cover for the MK 19 is an auxiliary component not common to all MK 64 mounts.

CHAPTER 4

MARKSMANSHIP

This chapter gives guidance for instruction and development of MK 19 gunnery skills. It provides data needed to develop training programs, plans, and lessons used to build the gunner's confidence and skill in firing the MK 19. Marksmanship training ensures that each soldier keeps his combat firing skills at a consistent level with the unit's mission. See Appendix D, Gunnery Tests, for more details concerning diagnostic and intermediate tests.

Section I. PREPARATORY MARKSMANSHIP TRAINING

Preparatory marksmanship training for the MK 19 covers the firing positions, the MK 19 fighting position, use of the T&E mechanism, and manipulation exercises. Thorough instruction during the preparatory training and exercises helps ensure efficient use of time and ammunition during range firing.

4-1. FIRING POSITIONS

Use the basic sitting, standing, and kneeling positions during training and range firing of the MK 19.

a. **Sitting.** When the tripod is used in the low or high position, sit directly behind the gun between the trail legs of the tripod. Extend your legs under the tripod, cross them, or brace your feet on the tripod (Figure 4-1). Place your elbows on the inside of your thighs for support when crossing your legs or bracing your feet on the tripod.

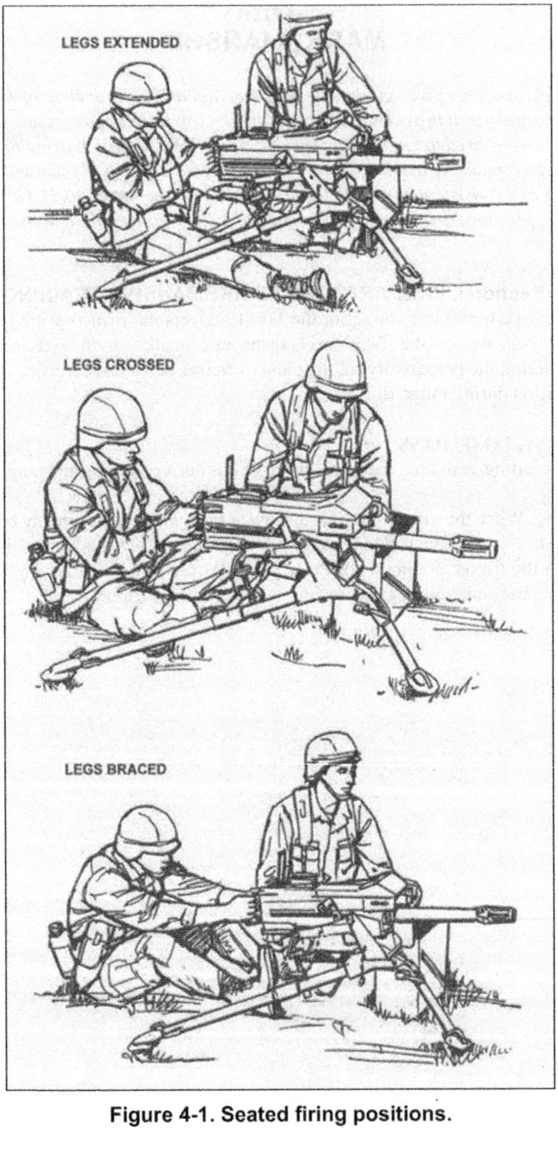

Figure 4-1. Seated firing positions.

b. **Standing.** When the MK 19 is mounted on a vehicle, stand with both of your hands on the control grips and your thumbs resting on the trigger. Keep your elbows against your body, your body forward, and your chest against your hands to brace the gun (Figure 4-2).

Figure 4-2. Standing position.

c. **Kneeling.** When the MK 19 is in a fighting or hasty tripod-mounted position, kneel and grasp the control grips with your thumbs on the trigger (Figure 4-3).

Figure 4-3. Hasty tripod position.

4-2. FIGHTING POSITION

A fighting position is dug when the unit has time to prepare it or is in a defensive position. Make the fighting position according to Figure 4-4.

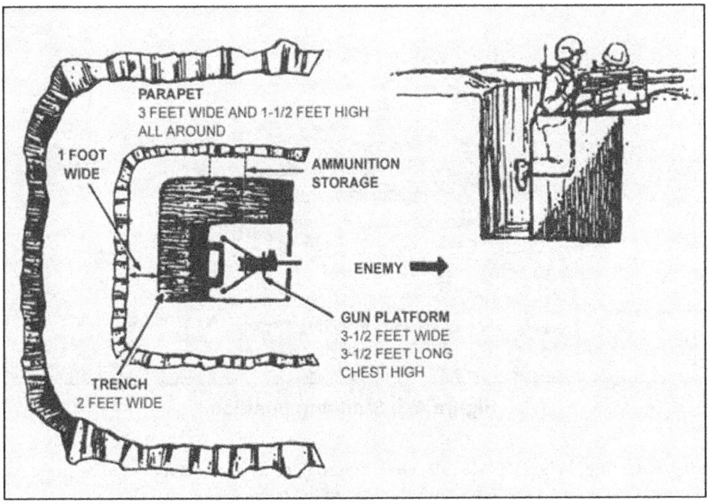

Figure 4-4. Dug-in position.

a. Position the MK 19 so it is oriented on the center of the assigned sector of fire. Mark the position of the tripod legs and trace the outline of the firing platform, which will be dug about 3 inches deep.

b. Dig the position in an "L" shape around the firing platform. The position should be chest-deep and wide enough to allow the MK 19 crew to load, operate, and place effective fire on an assigned sector of fire. The position can be made so that the gunner can fire from the kneeling position by digging the desired depth directly behind the gun.

c. Use the dirt from the hole to build flank parapets 3 feet wide and 1 ½ feet high. Dig the ammunition storage area into the left wall of the "L" and close to the ground.

d. Make sure the wall behind the firing platform is sloped to allow for entrance to and exit from the fighting position.

e. Follow these steps to build overhead cover:

(1) Put flank support logs, 4 to 6 inches in diameter, on top of each other along the entire length of the flank parapets.

(2) Put logs, 4 to 6 inches in diameter, side by side across the support logs as the base for the overhead cover.

(3) Put a waterproof layer over the base logs.

(4) Put 18 inches of dirt on top of the waterproofing material.

(5) Mold and camouflage the cover to blend with the terrain.

f. Camouflage the position with natural materials such as rocks, logs, live bushes, and grass.

4-3. TRAVERSE BAR AND T&E MECHANISM

Use the T&E mechanism to engage preselected target areas at night or during degraded light conditions. Record direction and elevation readings from the traversing bar and T&E mechanism. Record all readings in mils.

a. **Zero the T&E Mechanism.** Before the gunner can use the T&E mechanism to engage targets, he must zero it (Figure 4-5). Refer to section 3-2 for more information on zeroing the T&E mechanism.

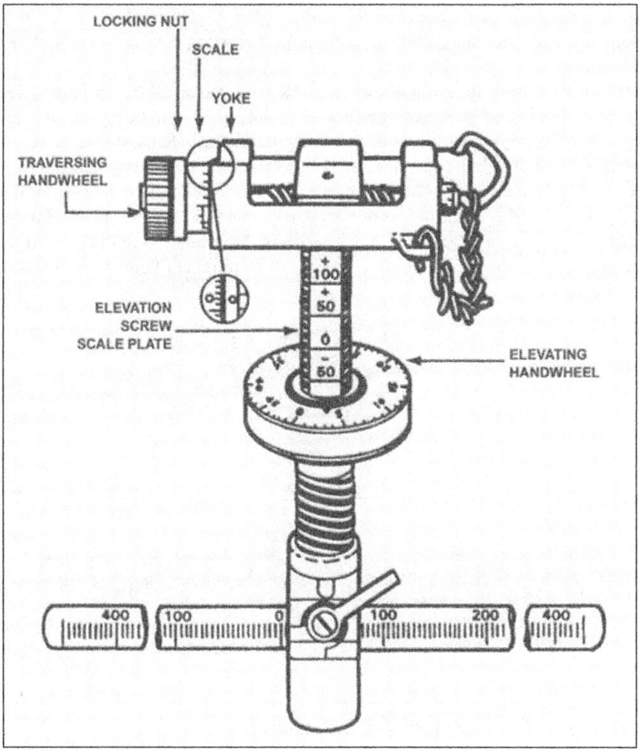

Figure 4-5. Zeroing the T&E mechanism.

b. **Lay the Gun for Direction.** When a sector of fire has been assigned, pick up the rear legs of the tripod and shift the tripod until the muzzle of the weapon points to the center of the sector of fire. Once the MK 19 is laid for direction, firmly stamp in the tripod shoes and place sandbags on the legs. This aids stability and may prevent accidental movement.

c. **Obtain and Record Direction Readings to all Targets Within the Sector of Fire** (Figure 4-6). Loosen the traversing slide lock lever and move the slide along the traversing bar until the MK 19 is laid on the center of a point target or on either flank of a linear target. Lock the traversing bar and read the direction from the scale on the traversing bar. If the left edge of the traversing slide does not fall exactly on a 5-mil tickmark, move the left edge of the traversing bar slide back to the next smaller mil reading. Use the traversing handwheel to complete the initial lay. Obtain a reading to the target by the direction of the MK 19 barrel. If the barrel is moved to the right, record a right reading. Read the number on the traversing bar from the left side of the traversing slide lock. If the barrel is moved to the left and the traversing slide lock is on the right side of the zero, record a left reading. After taking a direction reading for a target, measure the width of the target in mils using the traversing handwheel to move across the target. Reposition the traversing mechanism before moving to another target.

d. **Obtain Elevation Reading.** Ensure the MK 19 is laid on the center base of the target. Read the elevation from two scales: the first, or major, part from the elevating screw plate scale; the second, or minor, part from the elevating handwheel. Separate the two parts of the elevation reading with a slash (for example, -50/3). An elevation reading may not be valid on any other T&E mechanism than the one from which it was read. If the number of threads is increased or decreased after the data is recorded, accurate fire cannot be placed on the target. For example, if a weapon is freed to engage a secondary sector, and the base of the T&E mechanism is rotated, the data is wrong unless the same number of threads is exposed before and after the move.

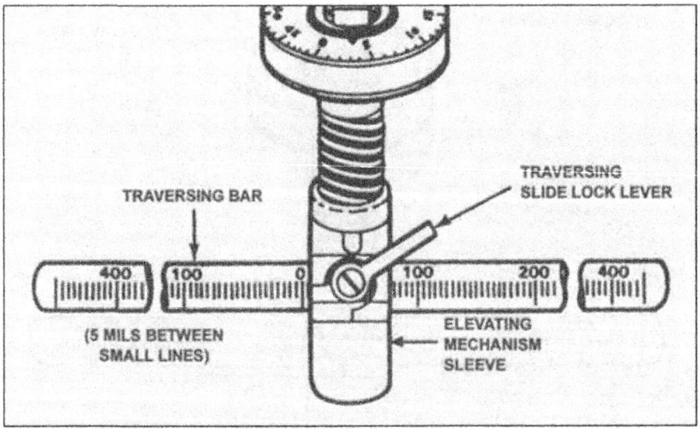

Figure 4-6. Direction reading.

(1) ***Major Reading.*** The elevating scale plate is graduated in 50-mil increments from -250 to +100. Except for the zero, each number has a plus or minus sign above it. All numbers, including zero, have an index line below them. To get the major elevation reading, hold the

T&E mechanism so that the eyes are even with the top of the elevating handwheel. The major reading (for example, -50) is the first number, with an index line, that is visible above the handwheel (Figure 4-7).

(2) ***Minor Reading.*** The elevating handwheel is marked in 50 increments of 1 mil each (Figure 4-7). The minor reading, which has no plus or minus sign, is the number that lines up with the pointer. The minor reading represents the number of mils the gun is laid below the major reading, for example, 3 mils. Verify a direction elevation reading by firing and adjusting on the target. Use the dry-fire method to get data to targets without live firing and adjusting. Set the rear sight on the range to the target, lay the gun on the center base of the target, and take direction and elevation readings. Use the dry-fire method only when firing is not feasible.

NOTE: Correct range setting is critical since any discrepancy causes an error in the elevation when the target is engaged.

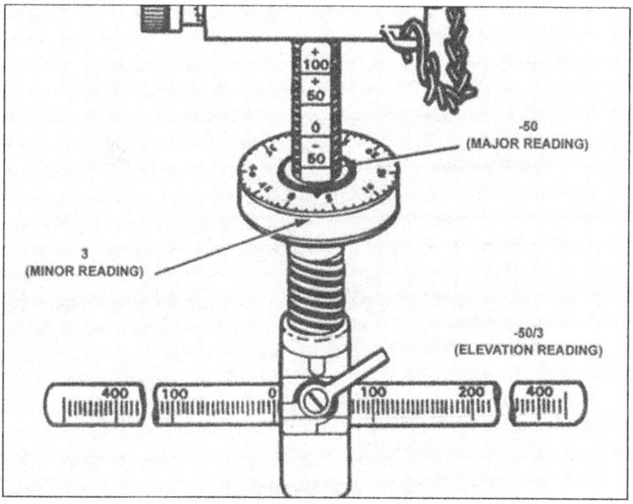

Figure 4-7. Elevation readings.

4-4. EXERCISES

The exercises include sighting and aiming; sight setting and laying; manipulating; determining the range; and preparing a range card.

a. **Sighting and Aiming Exercises.** There are two sighting and aiming exercises. The first calls for the use of the sighting bar.

(1) This exercise teaches how to obtain the correct sight picture, and also teaches alignment methods. Sight alignment means lining up the front and rear sights, with the top center of the front sight blade appearing in the center of the rear sight. Sight picture means lining up the front sight, the rear sight, and the target, with the tip of the front sight blade just touching the bottom center of the target. Gunners must show skill in aligning sights correctly and in getting the correct sight picture, using a sighting bar (Figure 4-8). Gunners must correctly show the point of aim, after sight alignment, 8 out of 10 times.

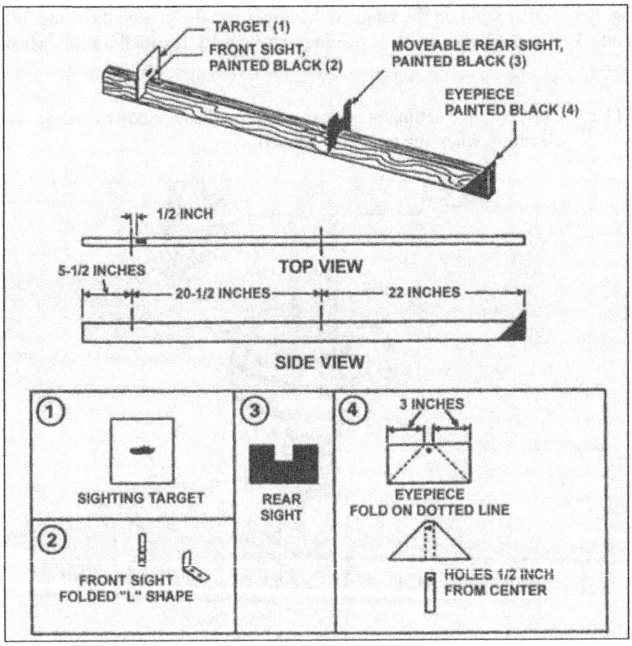

Figure 4-8. Sighting bar.

(2) The second sighting and aiming exercise, used with the MK 19 and target board, allows the gunner to apply lessons learned during the first exercise. Place the target board 10 meters from the muzzle of the MK 19 (Figure 4-9). Use the T&E mechanism to get the correct sight picture on each target that is called out. When the gunner feels he can sight and aim, the coach checks and critiques the gunner's ability to change from the start point on the target board to any new target called by the coach.

NOTE: Continue this exercise until the gunner is skilled in sighting and aiming.

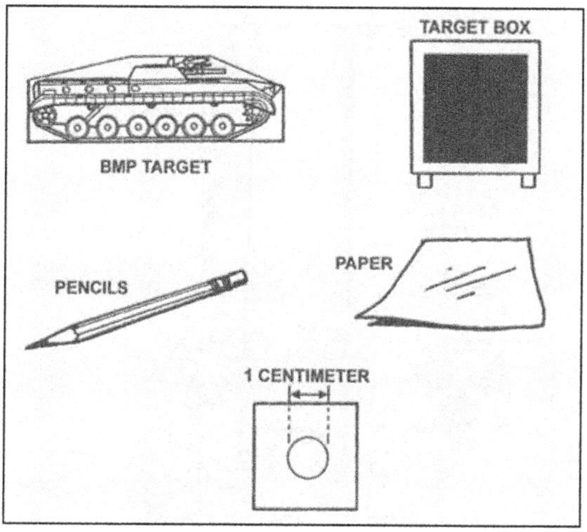

Figure 4-9. MK 19 target board.

b. **Sight Setting and Laying Exercises.** This first exercise teaches the soldier how to set the rear sight quickly and correctly (Figure 4-10). The second builds accuracy and speed in laying the gun on an aiming point, and allows extra practice in sight setting. Explain and show each exercise before beginning instruction.

(1) ***Sight Setting Exercise.*** Explain that, by training the elevating screw knob, the soldier can make minor adjustments in elevation. All major adjustments are made by using the slide release. Have one soldier act as gunner and another act as coach. Announce the range (for example, "EIGHT HUNDRED"). Have the soldier in the gunner's position conduct the following:

- Repeat the range.
- Set the sight at the announced range.
- Assume the correct gunner's position.
- Announce "Up."

The soldier, acting as coach, checks the setting of the slide and points out any errors. The soldiers work in pairs, change roles, and go through the exercise until each one can correctly and rapidly set the sight.

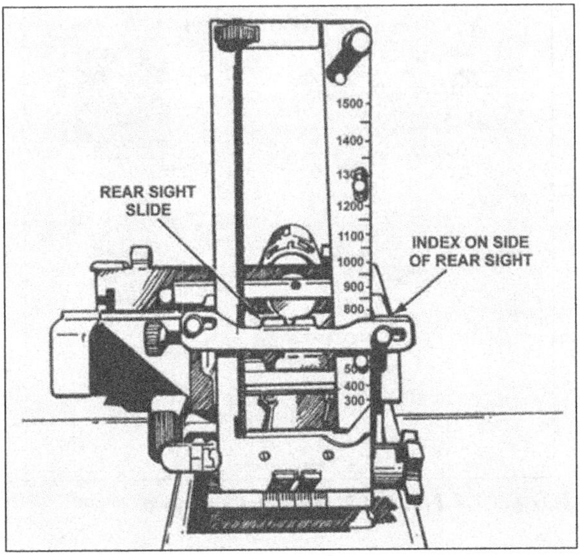

Figure 4-10. Rear sight adjustments.

(2) ***Laying Exercise.*** The MK 19 target board is used in this exercise. Explain that the exercise starts with the sight leaf down and the sight slide at 500 meters. Ranges less than 400 meters or greater than 1,500 meters are not announced. Have one soldier act as the gunner and another act as the coach at the gun. Announce an aiming point and range (for example, "TARGET NUMBER ONE," "ONE TWO HUNDRED"). Pause after each element to allow the gunner time to repeat it. After the gunner repeats each element, have the gunner:

- Set the sights
- Manipulate the gun by turning the T&E handwheels until sights are aligned on the chosen aiming point.
- Assume the correct gunner's position.
- Report "Up."

The soldier, acting as coach, checks the sight setting and lay of the gun. The soldiers work in pairs, changing roles until each is skilled in doing the exercise.

c. **Manipulation Exercise.** Manipulation means shifting the direction of the gun from one point to another. There are two manipulation exercises.

(1) The first exercise gives the gunner practice in manipulation. The coach stands 10 paces in front of the gun and uses hand signals to show the direction in which the gunner is to move the muzzle (Figure 4-11). The gunner works the handwheel to manipulate the gun. The coach observes and makes needed corrections. When the gunner reacts quickly and can manipulate the gun as he has been shown, his instruction may continue.

Figure 4-11. Coach and gunner exercise.

NOTE: Continue this exercise until the gunner is skilled in sighting and aiming.

(2) Once the soldier knows the basics of sighting and aiming, and can assume a good firing position, give him the second manipulation exercise. Show him how to manipulate the weapon to get a correct initial lay and how to skillfully shift the direction of the weapon to successive points. Ensure the soldier understands and knows how to perform the following instructions:

(a) Place the MK 19 target board 10 meters from the muzzle of the gun.

(b) Make large shifts in direction by releasing the traversing slide lock lever and moving the slide to the right or left. Make small changes in direction by turning the traversing handwheel with the left hand. One click on either T&E handwheel moves the strike 1 mil (1 cm on the target).

(c) Manipulate for elevation by rotating the elevating handwheel with the left hand (Figure 4-12).

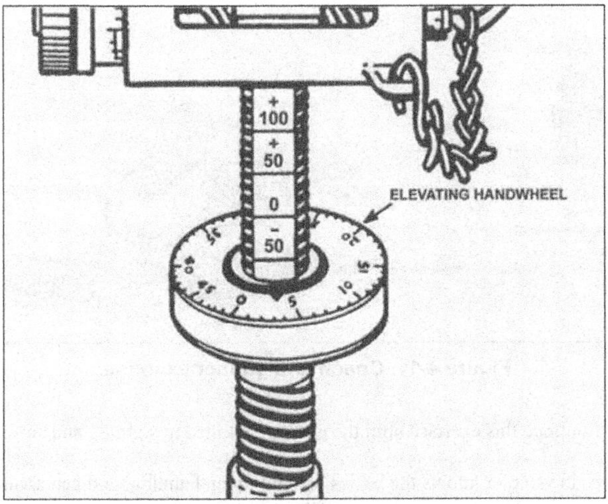

Figure 4-12. Elevating handwheel.

d. **Range Determination Exercise.** Range determination means finding the distance between the gunner's position and his target. The gunner's ability to engage a target depends on his ability to correctly find the range to the target. Under combat conditions, ranges may not be known in advance; the effectiveness of fire depends largely on the accuracy of the range used. There are several methods for finding range. Some of these methods include: estimating visually, firing the weapon, measuring range from a map or aerial photograph, and pacing the distance. Binoculars or laser range finders may be used (see Chapter 5, Techniques of Fire, for more on range determination).

e. **Range Card Exercise.** A range card is a record of the firing data needed to engage predetermined targets within a sector of fire at night or during degraded conditions. The range card may also be used to aid in target engagement during good visibility. It aids the leader in preparing his defense plan. Predetermined targets in the secondary sector are engaged by use of field expedients. A range card has two parts: a sketch section and a data section. Although the sketch is not drawn to scale, the data referring to the targets must be correct. DA Form 5517-R is the standard range card to be used (Figure 4-13).

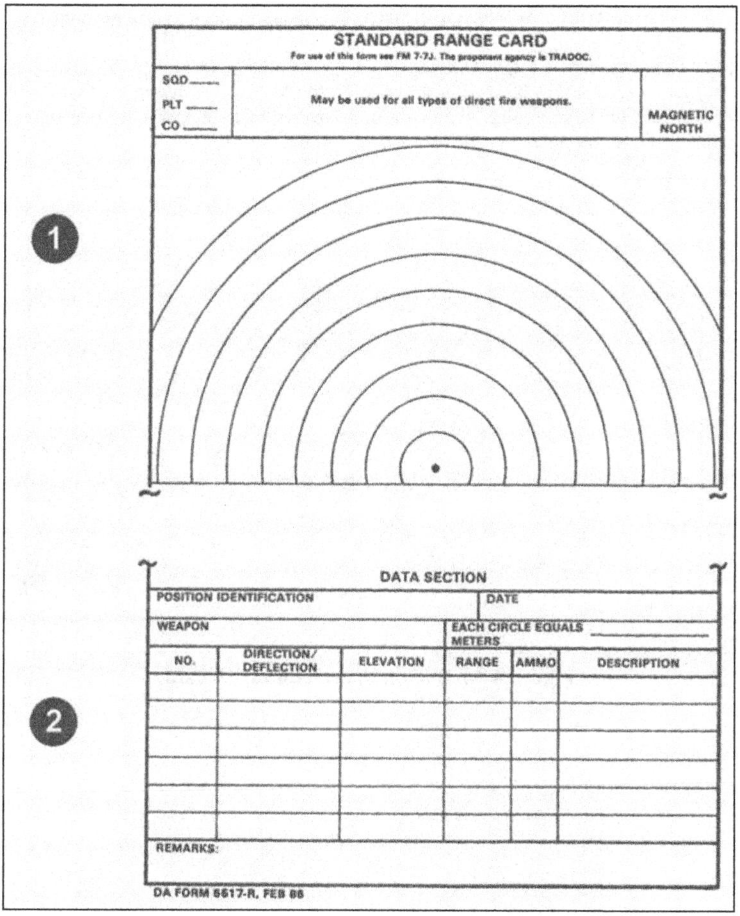

Figure 4-13. Range card sketch (1) and data sections (2).

(1) Range cards are prepared for primary, alternate, and supplementary positions. They are prepared immediately upon arrival in a position, regardless of the length of stay, and are updated as necessary. Range cards are prepared in duplicate. One copy stays at the gun position, and the other is sent to headquarters. Range cards are prepared on arrival in a position, regardless of the planned length of stay. Range cards are revised often during occupation of a position.

(2) Once the leader has decided where the firing position will be, the gunner prepares the range card.

(3) The range card is prepared as follows:

(a) Draw a sector sketch, covering the entire sector. Make the sketch as large as possible, not to exceed the largest circle. For a large area covered by trees or woods, draw only the outline and label the area "woods" or "orchard" (Figure 4-14).

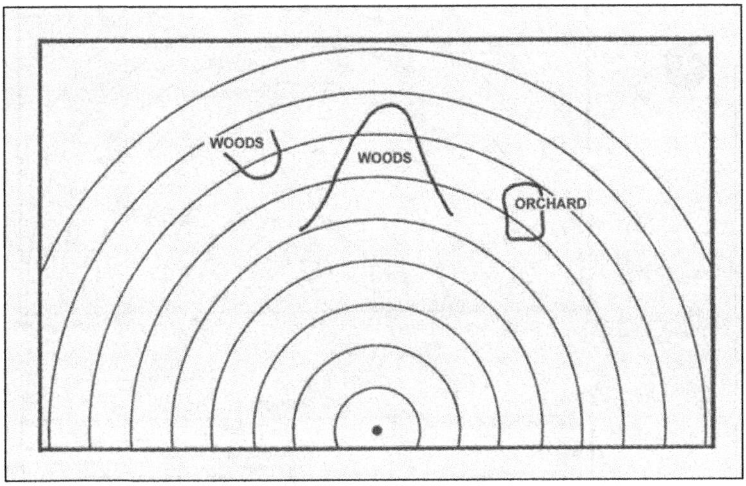

Figure 4-14. Sketch of area on range card.

(b) In the lower center of the sketch section, show the firing position by drawing the symbol for the weapon (Figure 4-15).

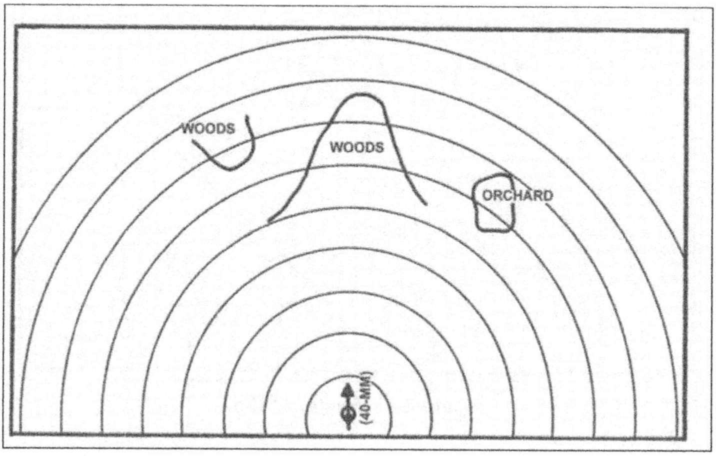

Figure 4-15. MK 19 symbol on range card.

(c) Show the location of the firing position by drawing a sketch of a nearby recognizable terrain feature. Label it and draw an arrow to the weapon symbol. Add the distance and azimuth from the terrain feature to the firing position (Figure 4-16).

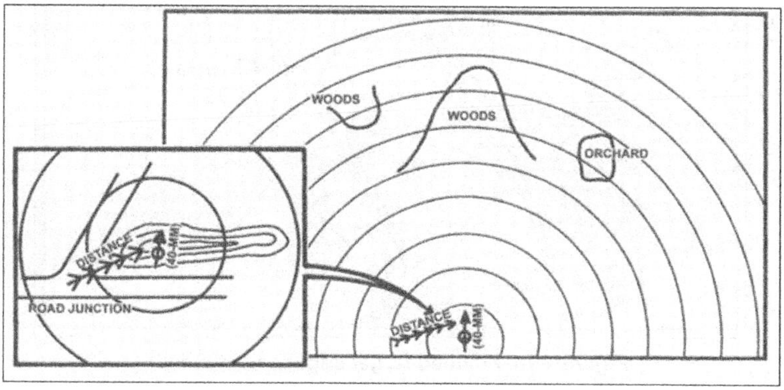

Figure 4-16. Distance and azimuth to terrain feature.

(d) Draw lines from the weapon symbol to reflect the left and right limits (Figure 4-17).

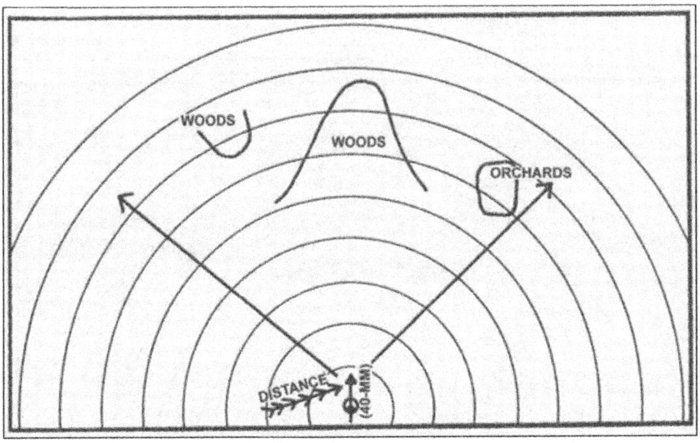

Figure 4-17. Sector of fire.

(e) Number the planned target engagement locations from left to right, and write them on the range card (Figure 4-18).

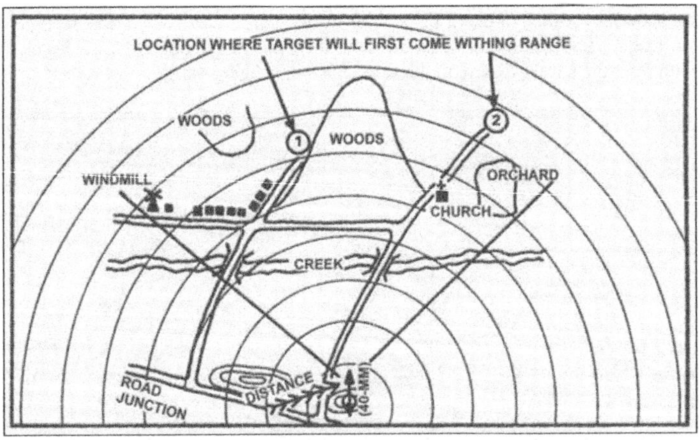

Figure 4-18. Planned target engagement areas.

(f) Number the target reference points (TRPs) and write them on the range card (Figure 4-19).

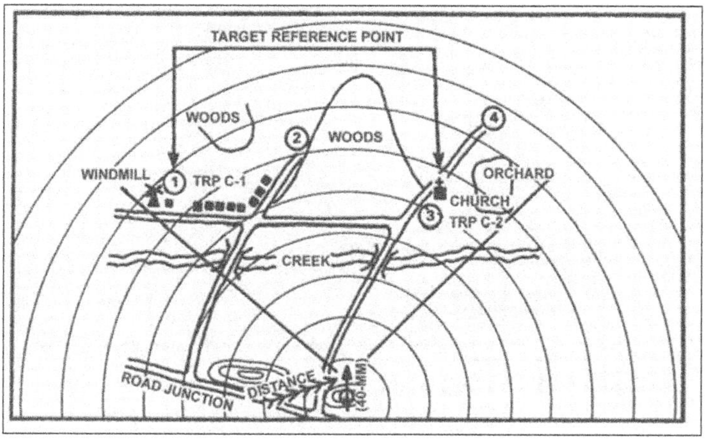

Figure 4-19. Target reference points.

(g) If no limiting factors exist, the maximum engagement line curves and joins the left and right sector-of-fire boundaries at the maximum engagement range (Figure 4-20). If limiting factors exist, draw the maximum engagement line in front of the limiting terrain features (Figure 4-21).

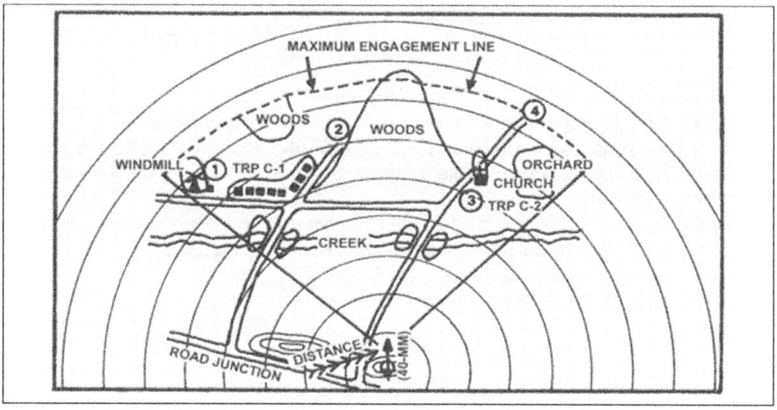

Figure 4-20. Maximum engagement line.

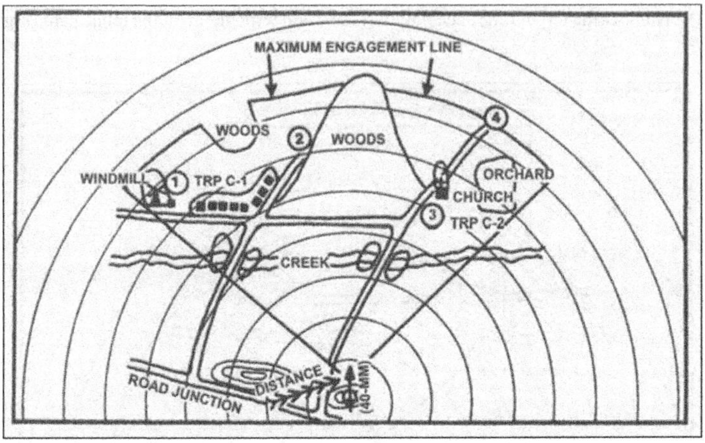

Figure 4-21. Limiting factors on maximum engagement line.

(h) Fill in the marginal information at the top of the card in the following manner (Figure 4-22):

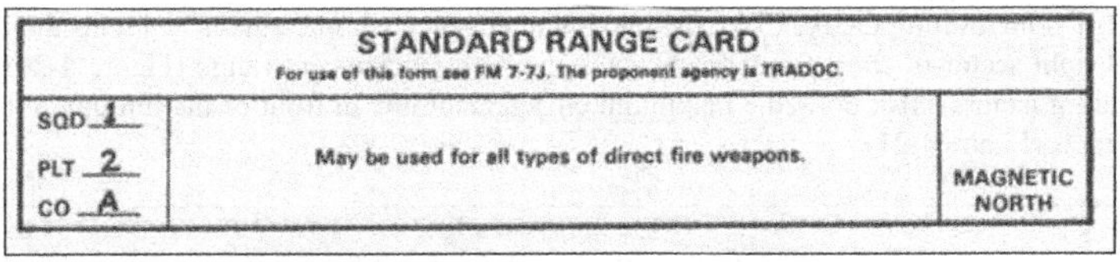

Figure 4-22. Marginal information.

- **Unit description.** Never show unit higher than company level.
- **Magnetic north.** Orient the range card with the terrain and find the direction of magnetic north with a compass. Draw a magnetic north arrow using the straight edge of the compass.

(i) Fill in the data section in the following manner (Figure 4-23):

- **Position identification.** Write primary, alternate, or secondary.
- **Weapon.** Write MK 19, grenade machine gun.
- **Date.** Write the day and month.

POSITION IDENTIFICATION PRIMARY			DATE 16 DEC		
WEAPON MK19			EACH CIRCLE EQUALS ______ METERS		
NO.	DIRECTION/ DEFLECTION	ELEVATION	RANGE	AMMO	DESCRIPTION
1	L200 MIL	+140/35	2050	M450	WINDMILL/TRP C-1
2	L30 M	+160/35	1750	M430	ROAD
3	R25 M	+140/35	1600	M430	CHURCH/TRP C-2
4	R100 M	+150/10	1400	M430	ROAD
REMARKS:					

Figure 4-23. Data section information.

- **Each circle equals 294 meters.** Write in the distance in meters between the circles. To find the distance, count the intervals from the weapon to the maximum engagement line. Divide the number of intervals into the range. This gives the distance between circles (Figure 4-24). For example:

$$\frac{2{,}212 \text{ meters}}{7.5 \text{ intervals}} = 294 \text{ meters between circles}$$

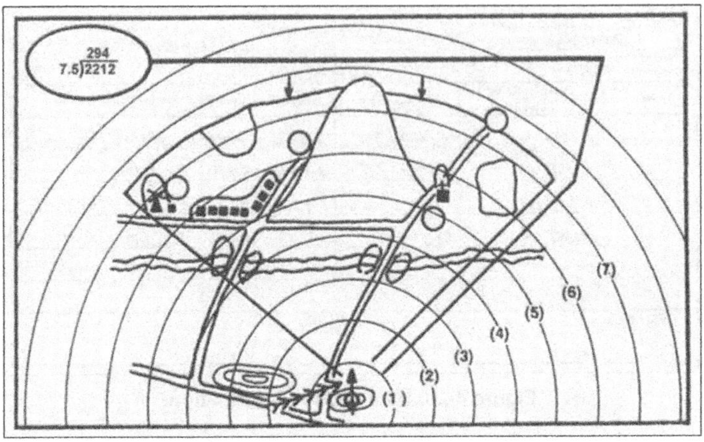

Figure 4-24. Determination of range the circles represent.

- **Number (No.).** Starting with number 1, write the numbers as listed for TRPs and target engagement locations.
- **Direction/deflection.** Write the mils from the traversing bar and handwheel.
- **Elevation.** Write the elevation from the elevating screw plate scale and the elevating handwheel.
- **Range.** Write distance in meters from the weapon to the TRP or target engagement area.
- **Ammunition.** Write the type of round issued for the mission.
- **Description.** List the name of the object; for example: road, windmill, or church. If the item is a TRP, list the TRP number also.

Section II. RANGE FIRE

Range firing gives hands-on MK 19 firing experience to the soldier who has completed the Intermediate Gunnery Test. Use the procedures in this section to set up and conduct range firing. The desired width for each lane is 100 meters at 800 meters from the firing line but can be modified due to local range or terrain limitations. Range firing includes zeroing procedures, observation and adjustment for fire, instructional firing, and qualification firing. Active fighting-force gunners qualify semiannually and participate in exercises and unit-level live fires quarterly. Active supporting-force gunners qualify and participate in instructional fire annually. All reserve component gunners qualify annually. See Appendix E, Annual Gunnery Training Program, for details.

4-5. DESCRIPTION

There are two types of ranges that may be used for MK 19 live-fire practice and qualification. Regardless of which range is used, the MK 19 can be fired from either a tripod or a vehicle.

a. **Multipurpose Gunnery Range for the 40-mm Grenade Machine Gun** (Figures 4-25 and 4-26). This range is designed for the conduct of individual firing exercises and qualifications against hull type targets. Details about this range can be found in TC 25-8.

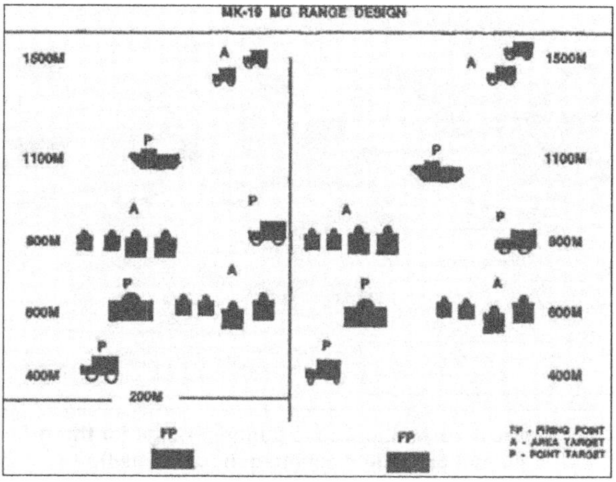

Figure 4-25. Multipurpose gunnery range for the 40-mm grenade machine gun.

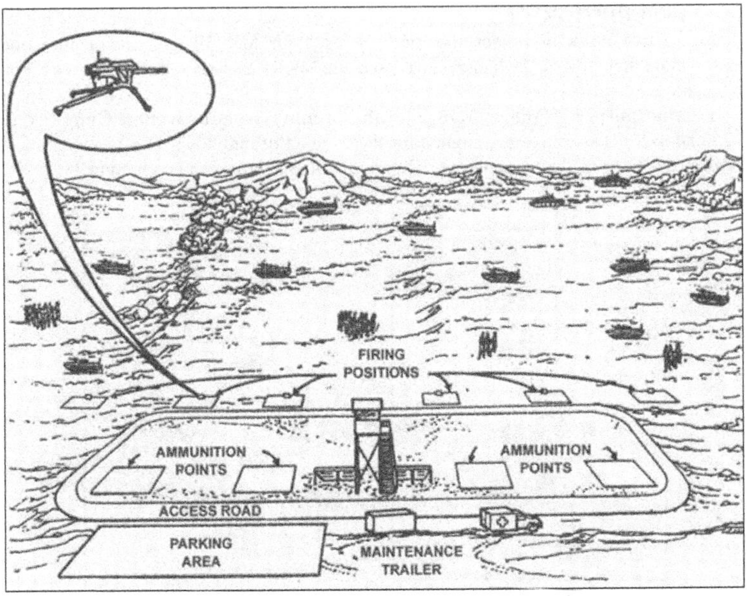

Figure 4-26. Multipurpose gunnery range for the 40-mm grenade machine gun (continued).

b. **Multipurpose Machine Gun Range.** The MPMG is the range suggested for MK 19 live-fire training. It is used to train mechanized, motorized, combat support, and combat service support units. The design and traits of this type of range are described in TC 25-8. Only a non dud-producing practice round is used on this range.

4-6. OPERATION

Use range operation procedures any time LFXs are conducted on the range. These procedures are the same for all range firing exercises.

a. Brief the gun crews on the range setup, fire commands, and safety procedures. Explain exercise needs as defined by the unit commander, or as they are to be done by the firing tables.

b. Divide the crews into groups and assign each group a firing point and an assistant instructor (AI). Ensure they are directed to their firing points and are shown their assigned sectors of fire. Have the AIs ensure that all crews have the equipment needed to perform the task.

c. Conduct the exercise. For example you can command GUNNER, WHEN I SAY BEGIN, PLACE THE MK 19 INTO OPERATION ON THE TRIPOD (ORGANIC CARRIER) WITHIN ____ MINUTES. ARE THERE ANY QUESTIONS? (pause) BEGIN.

At the end of the given time period, direct STOP. AIs, CRITIQUE YOUR CREWS AND GIVE ME YOUR "UP" SIGNAL WHEN YOU HAVE FINISHED.

4-7. COACHES

The instructor may arrange for a coach to be present at each gun during the following preparatory training, exercises, and instructional firing. Successful marksmanship training depends on the coaches. Well-trained and well-rehearsed gunners or crew members may be assigned as coaches. The coaches:

- Require the gunner to inspect his gun and equipment.
- Help the gunner place the gun on target.
- Require the gunner to explain the exercise that he is about to perform.
- Observe the gunner's position, grip, and manipulation during simulated firing.

4-8. FIRE PROCEDURES

Firing the MK 19 is not hard. However, to be effective, the gunner must be able to skillfully manipulate the trigger and T&E mechanism; the weapon must be held down and to one side while aiming and adjusting the T&E mechanism. The following steps are simple, but the gunner must remember to estimate the distance to the target, set the sights for distance, manipulate the T&E mechanism, press the trigger, and fire a single round; since the MK 19 is a fully automatic weapon, this can only be done by "popping" the trigger once. He must spot the impact of the first round and adjust to the target area. He should begin firing six- to nine-round bursts. This whole process can be done in a matter of seconds, the longest part being the flight time of the round. To fire the weapon:

(a) Place the safety on F (FIRE).
(b) Ensure the charger handles are in the forward and upright position.
(c) Place your hands on the control grips, and your thumb(s) on the trigger.
(d) Hold the weapon down and to one side, and check sight picture.
(e) Press the trigger to fire.

WARNING

Because of the gun's recoil, the first burst is the hardest to control. Ensure that the gun's recoil does not cause the barrel of the weapon to drop and rounds to strike short of the target. Keep it elevated.

4-9. ZEROING PROCEDURES

Zeroing procedures are crucial for hitting targets at ranges of 600 meters or more. It is strongly recommended that a target at 400 meters be used to zero. The following is the correct way to zero the MK 19 to the gunner.

a. Loosen the range plate screw. Move the range plate to the midpoint between the two studs. Tighten the range plate screw. Move the rear sight slide to the meter mark that represents the distance to the target. For example, move to the 400-meter mark to zero on a target known to be 400 meters away. Set the windage knob at the zero index line.

b. Align the sights on the base of the target using the T&E mechanism.

c. Fire a single round and spot the impact of the round. If it is on the target, fire another short burst to confirm the zero. If the round is not on target, go on to the next step.

d. To adjust for a round that is not on target, do the following:

(1) If the impact of the round is short or over, adjust the elevation knob. Estimate how short or over the round is. If the round is short, use this estimate to adjust the elevation knob clockwise, which moves the sights up onto the knob counterclockwise to bring the sights down to the target. For example, if the rounds impact 10 mils short, adjust the elevation knob 10 mils up by turning it clockwise. Realign the sights and adjust the gun back on target using the T&E mechanism before the next round is fired.

NOTE: 1. If the adjustment was correct, the second round should be on target. If so, fire the rest of the rounds to confirm the zero. If not, repeat the previous step.
2. If the impacts are not observed, bold adjustments may be necessary.

(2) If the round is to the right or left, adjust the windage knob. Estimate how far to the right or left the sight needs to move to bring the rounds on target. Turn the windage knob clockwise to adjust to the right; turn the windage knob counterclockwise to adjust to the left. For example, if the rounds impact 10 mils to the right, adjust the sight 10 mils to the left by turning the windage knob counterclockwise. Realign and adjust the sights back on target using the T&E mechanism before firing the next round.

(3) Once the zero is completed, align the range plate scale at the exact range of the target used to zero, and tighten it.

e. Point out errors and explain their effect.

f. When the gunner maintains the same sight picture, the type of firing position does not alter the zero.

4-10. OBSERVATION AND ADJUSTMENT OF FIRE

Observation and adjustment gives the soldier a chance to practice adjustment of fire by observing trajectory and impact or by re-laying often on the target using the sights. Live-fire training is held on the firing range after the methods are explained in the classroom.

a. When firing on targets, adjust by moving the burst into the target. Based on the strike of the rounds, adjust the number of clicks for elevation and direction needed to move the center of impact onto the target. This does not call for the use of sights. For example, fire at a target 500 meters away; the rounds impact 20 meters short and 50 meters right. Use the T&E handwheels, moving the muzzle left and up the proper number of clicks, to manipulate the gun onto the target.

b. Another procedure is the adjusted aiming point method. Use the sights and select an aiming point calculated to place the second burst on target. For example, fire at a 500-meter target; the rounds impact 20 meters short and 10 meters right. Rapidly select another aiming point about 20 meters beyond and 10 meters to the left of the target, lay on that aiming point, and fire.

4-11. INDIVIDUAL GUNNER EXERCISES AND QUALIFICATION

The individual gunnery exercises train and qualify MK 19 gunners. There are four scorecards available and they are used based on the type of target (hull or pop-up silhouettes) and whether the practice-qualification is during the day or during limited visibility. Each scorecard has two tables, one for practice and one for qualification. The tables have versions for hull or pop-up silhouette engagements and for the type of NVD used. Sample scorecards are shown in Figures 4-27 through 4-30 (and reproducible scorecards are provided in the back of this manual). MK 19s will be mounted or in the tripod configuration based on the range constraints and the commander's guidance. These tables are *recommended* tables for the infantry MK 19 gunner and crew. The first task in each table is a field zeroing evaluation, which allows the gunner to ensure his weapon is zeroed (even if he boresighted the MK 19). If the gunner fails to zero within four rounds, he is removed from the line and given additional training before attempting the table again. Refer to Section 4-5 for more detailed description of the ranges. The rest of the table consists of firing at individual and multiple targets.

a. **Day and Night Practice and Qualification.** Specific scorecards have been developed for different targets and NVDs. Gunners will only fire one day practice-qualification and one night practice-qualification. Units should select the practice and qualification based on the light conditions, type of targets available, and type of NVDs used. The following table shows which scorecards would be used:

Conditions	Target	Night Vision Devices	Scorecard
Day	Hull	NA	Scorecard I
Day	Pop-Up	NA	Scorecard III
Limited/Night	Hull	**AN/PEQ-2A** mounted on the TWS mounting bracket. **AN/PAS-13** mounted on the TWS mounting bracket. **AN/TVS-5** upgraded with the 3d generation tube mounted on the TWS mounting bracket.	Scorecard II
Limited/Night	Pop-Up or E-Type	All night vision devices.	Scorecard IV
Limited/Night	Any type	No night vision device.	Scorecard IV

Table 4-1. Scorecard matrix.

NOTE: 1. Both the MPMG range and the multipurpose gunnery range (MK 19) can be used for practice and qualification.

2. The MPMG range, modified with stationary armor targets at the required ranges, can be used as pop-up targets if a non dud-producing practice round is used. Scorecards III and IV are used on the MPMG range.

3. The multipurpose gunnery range (MK 19) can be used with any round authorized for the range. Scorecards I and II are used on the multipurpose gunnery range (MK 19).

Every target listed in each task is a point target; however one or two targets between the range of 600 and 900 meters should be changed to an area target. The number of rounds and engagement times should be the same for both point and area targets. The commander's guidance and the local range configuration should determine the location of the area targets.

(1) ***Day Practice and Qualification.*** Due to the types of targets available for practice and qualification, there are two scorecards for day practice and qualification. One (Table I) is used when engaging hull-type targets and the other (Table III) is used when engaging pop-up silhouette targets.

(a) *Hull-Type Targets*. Hull-type targets provide height, width, and depth, and give the MK 19 gunner a realistic target. The engagement ranges for practice and qualifications can therefore be set for the full range of the gun and are set at ranges up to 1,500 meters.

(b) *Pop-Up Silhouette Targets*. Pop-up silhouettes provide a target with width and height but very little depth. Due to the high angle of fall of 40-mm rounds at ranges greater than 800 meters, it is difficult to hit this type of target beyond that range. Therefore, the engagement ranges for practice and qualifications are set at 800 meters or less.

(c) *Engagement Times*. There is a 30 second difference for the completion of each task between the practice and qualification tables. Practice tables allow thirty additional seconds for each engagement.

b. **Night Practice and Qualification.** The MK 19 night practice and qualification tables are shown in Scorecards II and IV. Units with AN/PEQ-2A, AN/PAS-13, and AN/TVS-5 night aiming devices *and* engaging hull-type targets use Scorecard II. Units without a MK 19/sight combination or engaging pop-up silhouettes use Scorecard IV. Gunners do not fire both. Infantry gun crews are required to qualify at night. Other types of units may determine that day qualification is adequate due to their wartime missions.

c. **Scoring.** Scoring is done on a GO/NO GO basis for each task within the practice or qualification table.

(1) Zeroing the gun, the first task in each table, is scored as a GO/NO GO. Giving a score for the zero emphasizes the importance of a proper zero to effectively engage targets at 600 meters and beyond. However, if the gunner fails to zero within four rounds, he is removed from the line and given additional training before attempting the table again. This step reduces the waste of ammunition.

(2) On point target engagements (lightly armored vehicle targets such as BRDMs, threat scout cars, etc.), the gunner receives a GO if he meets or exceeds the engagement standard of one or two rounds hitting the target.

(3) If area targets are included (infantry squads, RPG teams, etc.), the gunner receives a GO when at least the number of rounds stated in the engagement standard for that task impact within ten meters of the area target and thus suppresses it.

(4) At the end of each table, the scorer adds up the number of GOs and NO GOs, places that number in the "Totals block, and checks the box to the left of the appropriate qualification (expert, sharpshooter, marksman, or unqualified).

d. **Range Setup.** Targets should be within the ranges provided on the table scorecard.

(1) Because the gunner has to be able to observe the impact of the round to make adjustments, there should be no dead space within 100 meters of the selected targets.

(2) Every target listed in each task is a point target; however, one or two targets between the range of 600 and 900 meters should be changed to an area target. The number of rounds and engagement time will be the same for both point and area targets. The commander's guidance and the local range configuration should determine the location of the area targets.

(3) For area targets, multiple E type personnel targets may be placed on line or in wedge formations. Multiple personnel targets, indicating area targets, should not be more than 5 meters apart, and not extend more than 30 meters in width or 20 meters in depth.

(4) During night firing using hulls as targets, no modification to the target is needed to assist the gunner in identifying the target. If pop-up silhouette targets are used however, a thermal source is needed on each target to enable the gunner to acquire it with the thermal weapon sight (TWS) and a light source is needed on each target if the AN/TVS-5 is being used. The thermal source can be two chemical lights on targets between 400 meters and 600 meters and three chemical lights on targets over 600 meters.

e. **Grading.** One grader will be required at each firing point.

(1) ***Grading Equipment.*** During the day, the grader will need a set of binoculars. At night equipment will vary according to the type of range being used. With an impact range with hull targets, the grader will need a NVD (examples, the AN/PVS-14,7B with the 3X magnifier, the AN/TAS-4, or the AN/PAS-13 [heavy]) to observe the strike of the round. The same equipment is needed if the pop up targets do not provide feedback. The grader also needs the order in which targets are engaged and a means to provide the gunner with the range to the target for that particular firing point. The grader must be able to identify which target is to be engaged by using, for example, a range card including a diagram of the range with targets numbered and ranges listed.

(2) ***Start and End Time.*** Time will start when the target is exposed and the grader has provided the target range (the graders will provide all information before the target is exposed). If hull targets are used and exposed at all times, then the time will start once the grader has told the gunner which target to engage and provided the range to the target. Time ends when the time indicated for that task expires, the target has been successfully engaged, or the target is no longer exposed.

f. **Ammunition.** Ammunition is broken down by task. The assistant gunner places each belt in its order of use. The number of rounds authorized for each task will be the number of rounds per belt. For example, if there are ten engagements, there should be ten belts of ammunition placed within reach of the assistant gunner in the order they are to be fired.

(1) HE rounds cannot be fired at pop-up silhouette targets because the lift mechanism will be damaged.

(2) Training practice tracer (TPT) rounds can be fired at both types of targets.

(3) The impact of a HE round is much easier to see than that of the TPT round.

g. **Fire Control.** Controlling and observing a target engagement with the MK 19 is not a problem with a range set up with a firing lane for each firing point. Especially with hull targets however, each point will not have an individual firing lane. Some ranges must use the same target for more than one lane, which may be a potential problem for grading. The grader must be able to identify which round impact is from which firing point. This is especially true for the 400-meter target. To prevent this problem, ensure that only one gunner at a time engages each target. The order in which the targets are engaged can be changed to allow more than one gunner to fire at the same time. Engagement start times can also be staggered so that gunners can engage targets at different ranges at the same time. This requires a great deal of coordination and communication between the graders and the personnel controlling the range.

h. **Day Practice and Day Qualification Firing Exercise** (Figures 4-27 and 4-28). Table I contains the tables for day practice and day qualification for hull targets and Table III contains the tables for day practice and day qualification for pop-up targets. Other than a 30 second difference in the engagement times for each task, the practice and the qualification tables are the same. It is held twice a year, or as often as the commander feels is needed to maintain gunner proficiency

(1) The day practice firing exercise allows the gunner to fire on a range engaging hull or pop-up targets to test his skills before qualification firing.

(2) The qualification LFX tests skills practiced during day firing exercise. It is scored on time taken and target hits made based on the firing tables.

(3) During scorecard preparation the grader selects the correct scorecard (Table I for hull targets or Table III for pop-up targets) and enters the gunner's name, rank, and unit in blocks "1" through "3." He also fills in blocks "4" through "7" with the range name, the firing lane, his name, and the date.

(4) The grader positions himself so that he can observe both the gunner and the target. Once live fire commences, he:

- Observes and informs the gunner the strike of each round.
- Observes and records a GO or NO GO for each task.

(5) At the end of the practice, the grader sums the total of GO/NO GOs in the "Totals" block, checks the appropriate qualification in block "9," has the gunner sign the scorecard in block "10," and signs the card in block "11."

(6) During the qualification phase, the grader repeats the steps above by filling in the appropriate blocks, summing the scores, and assigning the correct qualification.

(7) The grader can use the comment section in either table to enter remarks such as the operation of the gun, condition of the targets, and weather conditions to name a few.

MK 19, 40-mm GRENADE MACHINE GUN, MOD 3 FIRING TABLE I
DAY PRACTICE AND QUALIFICATION WITH HULL TARGETS SCORECARD
For use of this form, see FM 3-22.27; the proponent agency is TRADOC.

PRIVACY ACT STATEMENT

AUTHORITY:	10 USC 3012(g)/Executive order 9397
PRINCIPAL PURPOSE:	To aid individual training on targets at various ranges.
ROUTINE USES:	To evaluate individual proficiency.
DISCLOSURE:	Voluntary. However, mass rating and recording require some tracking method.

1a. LAST NAME	1b. FIRST NAME	1c. MI	2. RANK	3. UNIT
PEVOSKI	RICHARD	M	SSG	1/26TH IN

TABLE I (A). DISMOUNTED AND MOUNTED DAY PRACTICE

4. RANGE	5. LANE	6. GRADER	7. DATE
DOBOL	5	JOULWAN	4 SEP 03

TASK	RANGE (Meters)	AMMO	TIME (Minutes)	ENGAGEMENT STANDARDS	GO	NO GO
ZERO	400	4	NA	2 ROUNDS HIT	✓	
2	1,100 (+/- 200)	8	2.5	2 ROUNDS HIT		✓
3	1,500 (+/- 200)	10	3.5	2 ROUNDS HIT		✓
4	600 (+/- 100)	6	2	2 ROUNDS HIT	✓	
5	800 (+/- 100)	6	2	2 ROUNDS HIT	✓	
6	400	4	1.5	2 ROUNDS HIT	✓	
MULTIPLE TARGETS						
7	1,100 (+/- 200)	10	4	1 ROUND HIT	✓	
8	600 (+/- 100)			1 ROUND HIT	✓	
9	800 (+/- 100)	14	4.5	1 ROUND HIT	✓	
10	1,500 (+/- 200)			1 ROUND HIT	✓	
				TOTALS	8	

8. COMMENTS: USED HE

9. NUMBER OF ENGAGEMENT MET *(Choose One)*
- [ ] 10 - EXPERT
- [ ] 9 - SHARPSHOOTER
- [x] 8-7 - MARKSMAN
- [ ] 6 AND BELOW - UNQUALIFIED

10. GUNNER'S SIGNATURE: Richard M. Pevoski
11. GRADER'S SIGNATURE: George Joulwan

TABLE I (B). DISMOUNTED AND MOUNTED DAY QUALIFICATION

12. RANGE	13. LANE	14. GRADER	15. DATE
DOBOL	8	JOULWAN	4 SEP 03

TASK	RANGE (Meters)	AMMO	TIME (Minutes)	ENGAGEMENT STANDARDS	GO	NO GO
ZERO	400	4	NA	2 ROUNDS HIT	✓	
2	1,100 (+/- 200)	8	2	2 ROUNDS HIT	✓	
3	1,500 (+/- 200)	10	3	2 ROUNDS HIT		✓
4	600 (+/- 100)	6	1.5	2 ROUNDS HIT	✓	
5	800 (+/- 100)	6	1.5	2 ROUNDS HIT	✓	
6	400	4	1	2 ROUNDS HIT	✓	
MULTIPLE TARGETS						
7	1,100 (+/- 200)	10	3.5	1 ROUND HIT	✓	
8	600 (+/- 100)			1 ROUND HIT	✓	
9	800 (+/- 100)	14	4	1 ROUND HIT	✓	
10	1,500 (+/- 200)			1 ROUND HIT	✓	
				TOTALS	9	

16. COMMENTS: USED HG

17. NUMBER OF ENGAGEMENT MET *(Choose One)*
- [ ] 10 - EXPERT
- [x] 9 - SHARPSHOOTER
- [ ] 8-7 - MARKSMAN
- [ ] 6 AND BELOW - UNQUALIFIED

18. GUNNER'S SIGNATURE: Richard M. Pevoski
19. GRADER'S SIGNATURE: George Joulwan

DA FORM 7518-R, AUG 2003 APD V1.00

Figure 4-27. Example of completed DA FORM 7518-R, MK 19 day practice and qualification (hull targets).

MK 19, 40-mm GRENADE MACHINE GUN, MOD 3 FIRING TABLE III
DAY PRACTICE AND QUALIFICATION WITH POP-UP TARGETS SCORECARD
For use of this form, see FM 3-22.27; the proponent agency is TRADOC.

PRIVACY ACT STATEMENT

AUTHORITY: 10 USC 3012(g)/Executive order 9397
PRINCIPAL PURPOSE: To aid individual training on targets at various ranges.
ROUTINE USES: To evaluate individual proficiency.
DISCLOSURE: Voluntary. However, mass rating and recording require some tracking method.

1a. LAST NAME	1b. FIRST NAME	1c. MI	2. RANK	3. UNIT
GUNNING	ROBERT	T	SGT	1/26TH IN

TABLE III (A). DISMOUNTED AND MOUNTED DAY PRACTICE

4. RANGE	5. LANE	6. GRADER	7. DATE
SEITZ	2	MULLENIX	9 SEP 03

TASK	RANGE (Meters)	AMMO	TIME (Minutes)	ENGAGEMENT STANDARDS	GO	NO GO
ZERO	400	4	NA	2 ROUNDS HIT	✓	
2	600 (+/- 100)	6	2	1 ROUND HIT	✓	
3	800 (+/- 100)	8	2.5	1 ROUND HIT	✓	
4	400	4	1.5	1 ROUND HIT	✓	
MULTIPLE TARGETS						
5	800 (+/- 100)	12	4	1 ROUND HIT		✓
6	400			1 ROUND HIT	✓	
7	400	10	3	1 ROUND HIT	✓	
8	600 (+/- 100)			1 ROUND HIT	✓	
				TOTALS	7	

8. COMMENTS: CHECK TARGET @ 800M

9. NUMBER OF ENGAGEMENT MET *(Choose One)*
- [ ] 8 - EXPERT
- [ ] 6 - MARKSMAN
- [x] 7 - SHARPSHOOTER
- [ ] 5 AND BELOW - UNQUALIFIED

10. GUNNER'S SIGNATURE: Robert T. Gunning
11. GRADER'S SIGNATURE: J. Mullenix

TABLE III (B). DISMOUNTED AND MOUNTED DAY QUALIFICATION

12. RANGE	13. LANE	14. GRADER	15. DATE
SEITZ	5	MULLENIX	9 SEP 03

TASK	RANGE (Meters)	AMMO	TIME (Minutes)	ENGAGEMENT STANDARDS	GO	NO GO
ZERO	400	4	NA	2 ROUNDS HIT	✓	
2	600 (+/- 100)	6	1.5	1 ROUND HIT	✓	
3	800 (+/- 100)	8	2	1 ROUND HIT	✓	
4	400	4	1	1 ROUND HIT	✓	
MULTIPLE TARGETS						
5	800 (+/- 100)	12	3.5	1 ROUND HIT		✓
6	400			1 ROUND HIT		✓
7	400	10	2.5	1 ROUND HIT	✓	
8	600 (+/- 100)			1 ROUND HIT	✓	
				TOTALS	6	

16. COMMENTS: T:6 CAME LOOSE
T 5:6

17. NUMBER OF ENGAGEMENT MET *(Choose One)*
- [ ] 8 - EXPERT
- [x] 6 - MARKSMAN
- [ ] 7 - SHARPSHOOTER
- [ ] 5 AND BELOW - UNQUALIFIED

18. GUNNER'S SIGNATURE: Robert T. Gunning
19. GRADER'S SIGNATURE: J. Mullenix

DA FORM 7520-R, AUG 2003 APD V1.00

Figure 4-28. Example of completed DA FORM 7520-R, MK 19 day practice and qualification (pop-up targets).

i. **Night Practice and Night Qualification Firing Exercises** (Figures 4-29 and 4-30). Table II contains the tables for night practice and night qualification for hull targets using the AN/PEQ-2A, AN/PAS-13, or the AN/TVS-5 NVDs. Table IV contains the tables for night practice and night qualification with pop-up targets or hull targets without using NVDs. Other than a 30 second difference in the engagement times for each task, the practice and the qualification tables are the same. It is held twice a year, or as often as the commander feels is needed to maintain gunner proficiency.

(1) The night practice firing exercise allows the gunner to fire on a range engaging hull or pop-up targets to test his skills before qualification firing.

(2) The qualification exercise tests skills practiced during night firing exercise. It is scored on time taken and target hits made based on the firing tables.

(3) The grader selects the correct scorecard (Table II for hull targets using the NVDs listed on the scorecard, or Table IV for pop-up targets or hull targets without a NVD) and enters the gunner's name, rank, and unit in blocks "1" through "3." He also fills in blocks "4" through "7" with the range name, the firing lane, his name, and the date. He also checks the appropriate NVD in block "8" (Table II only).

(4) The grader positions himself so that he can observe both the gunner and the target. Once live fire commences, he:

- Observes and informs the gunner the strike of each round.
- Observes and records a GO or NO GO for each task.

(5) At the end of the practice, the grader sums the total of GO/NO GOs in the "Totals" block , checks the appropriate qualification in block "9," has the gunner sign the scorecard in block "10," and signs the card in block "11."

(6) During the qualification phase, the grader repeats the steps above by filling in the appropriate blocks, summing the scores, and assigning the correct qualification.

(7) The grader can use the back of the form or the comment section to enter remarks such as the operation of the gun, condition of the targets, and weather conditions to name a few.

MK 19, 40-mm GRENADE MACHINE GUN, MOD 3 FIRING TABLE II
NIGHT PRACTICE AND QUALIFICATION WITH HULL TARGETS SCORECARD

For use of this form, see FM 3-22.27; the proponent agency is TRADOC.

PRIVACY ACT STATEMENT

AUTHORITY:	10 USC 3012(g)/Executive order 9397
PRINCIPAL PURPOSE:	To aid individual training on targets at various ranges.
ROUTINE USES:	To evaluate individual proficiency.
DISCLOSURE:	Voluntary. However, mass rating and recording require some tracking method.

NOTE: Use these tables if the targets being used are hulls, and any of the following applies:
AN/PEQ-2A mounted on the TWS mounting bracket.
AN/PAS-13 mounted on the TWS mounting bracket.
AN/TVS-5 with the 3d generation tube mounted on the TWS mounting bracket.

1a. LAST NAME	1b. FIRST NAME	1c. MI	2. RANK	3. UNIT
PEVOSKI	RICHARD	M	SSG	1/26 IN

TABLE II (A). DISMOUNTED AND MOUNTED NIGHT PRACTICE

4. RANGE	5. LANE	6. GRADER	7. DATE
DOBOL	2	CROSS	2 SEP 03

TASK	RANGE (Meters)	AMMO	TIME (Minutes)	ENGAGEMENT STANDARDS	GO	NO GO
ZERO	400	4	NA	2 ROUNDS HIT	✓	
2	1,100 (+/- 200)	8	2.5	2 ROUNDS HIT	✓	
3	1,500 (+/- 200)	10	3.5	2 ROUNDS HIT		✓
4	600 (+/- 100)	6	2	2 ROUNDS HIT	✓	
5	800 (+/- 100)	6	2	2 ROUNDS HIT		✓
6	400	4	1.5	2 ROUNDS HIT	✓	
MULTIPLE TARGETS						
7	1,100 (+/- 200)	10	4	1 ROUND HIT		✓
8	600 (+/- 100)			1 ROUND HIT	✓	
9	800 (+/- 100)	14	4.5	1 ROUND HIT	✓	
10	1,500 (+/- 200)			1 ROUND HIT		✓
				TOTALS	6	

8. TYPE DEVICE *(Choose One)*: ☐ AN/PEQ-2A ☐ AN/PAS-13 ☑ AN/TVS-5

9. NUMBER OF ENGAGEMENT MET *(Choose One)*: ☐ 10 - EXPERT ☐ 9 - SHARPSHOOTER ☐ 8-7 - MARKSMAN ☑ 6 AND BELOW - UNQUALIFIED

10. GUNNER'S SIGNATURE: Richard M. Pevoski

11. GRADER'S SIGNATURE: Arnie Cross

TABLE II (B). DISMOUNTED AND MOUNTED NIGHT QUALIFICATION

12. RANGE	13. LANE	14. GRADER	15. DATE
DOBOL	3	CROSS	2 SEP 03

TASK	RANGE (Meters)	AMMO	TIME (Minutes)	ENGAGEMENT STANDARDS	GO	NO GO
ZERO	400	4	NA	2 ROUNDS HIT	✓	
2	1,100 (+/- 200)	8	2	2 ROUNDS HIT	✓	
3	1,500 (+/- 200)	10	3	2 ROUNDS HIT		✓
4	600 (+/- 100)	6	1.5	2 ROUNDS HIT	✓	
5	800 (+/- 100)	6	1.5	2 ROUNDS HIT		✓
6	400	4	1	2 ROUNDS HIT	✓	
MULTIPLE TARGETS						
7	1,100 (+/- 200)	10	3.5	1 ROUND HIT		✓
8	600 (+/- 100)			1 ROUND HIT	✓	
9	800 (+/- 100)	14	4	1 ROUND HIT	✓	
10	1,500 (+/- 200)			1 ROUND HIT		✓
				TOTALS	6	

16. TYPE DEVICE *(Choose One)*: ☐ AN/PEQ-2A ☐ AN/PAS-13 ☑ AN/TVS-5

17. NUMBER OF ENGAGEMENT MET *(Choose One)*: ☐ 10 - EXPERT ☐ 9 - SHARPSHOOTER ☐ 8-7 - MARKSMAN ☑ 6 AND BELOW - UNQUALIFIED

18. GUNNER'S SIGNATURE: Richard M. Pevoski

19. GRADER'S SIGNATURE: Arnie Cross

DA FORM 7519-R, AUG 2003 APD V1.00

Figure 4-29. Example of completed DA FORM 7519-R, MK 19 night practice and qualification with hull targets and NVDs.

MK 19, 40-mm GRENADE MACHINE GUN, MOD 3 FIRING TABLE IV
NIGHT PRACTICE AND QUALIFICATION WITH POP-UP TARGETS SCORECARD
For use of this form, see FM 3-22.27; the proponent agency is TRADOC.

PRIVACY ACT STATEMENT

AUTHORITY: 10 USC 3012(g)/Executive order 9397
PRINCIPAL PURPOSE: To aid individual training on targets at various ranges.
ROUTINE USES: To evaluate individual proficiency.
DISCLOSURE: Voluntary. However, mass rating and recording require some tracking method.

NOTE: Use this table if you do not have a MK 19/sight combination that applies to Table II, and you are using hull targets OR if targets are pop-up/E type silhouettes.

1a. LAST NAME	1b. FIRST NAME	1c. MI	2. RANK	3. UNIT
GUNNING	ROBERT	T	SGT	1/26th IN

TABLE IV (A). DISMOUNTED AND MOUNTED NIGHT PRACTICE

4. RANGE	5. LANE	6. GRADER	7. DATE
SEITZ	2	MULLENIX	9 SEP 03

TASK	RANGE (Meters)	AMMO	TIME (Minutes)	ENGAGEMENT STANDARDS	GO	NO GO
ZERO	400	4	NA	2 ROUNDS HIT	✓	
2	600 (+/- 100)	6	2	2 ROUNDS HIT	✓	
3	800 (+/- 100)	6	2	2 ROUNDS HIT	✓	
4	400	4	1.5	2 ROUNDS HIT	✓	
MULTIPLE TARGETS						
5	800 (+/- 100)	10	3.5	1 ROUND HIT		✓
6	400			1 ROUND HIT		✓
7	400	14	2.5	1 ROUND HIT	✓	
8	600 (+/- 100)			1 ROUND HIT	✓	
				TOTALS	6	

8. COMMENTS
OVERCAST BUT TARGETS VISIBLE.

9. NUMBER OF ENGAGEMENT MET *(Choose One)*
☐ 8 - EXPERT
☑ 6 - MARKSMAN
☐ 7 - SHARPSHOOTER
☐ 5 AND BELOW - UNQUALIFIED

10. GUNNER'S SIGNATURE: Robert T. Gunning
11. GRADER'S SIGNATURE: J. Mullenix

TABLE IV (B). DISMOUNTED AND MOUNTED NIGHT QUALIFICATION

12. RANGE	13. LANE	14. GRADER	15. DATE
SEITZ	4	CORLEY	9 SEP 03

TASK	RANGE (Meters)	AMMO	TIME (Minutes)	ENGAGEMENT STANDARDS	GO	NO GO
ZERO	400	4	NA	2 ROUNDS HIT	✓	
2	600 (+/- 100)	6	1.5	2 ROUNDS HIT	✓	
3	800 (+/- 100)	6	1.5	2 ROUNDS HIT	✓	
4	400	4	1	2 ROUNDS HIT		✓
MULTIPLE TARGETS						
5	800 (+/- 100)	10	3	1 ROUND HIT	✓	
6	400			1 ROUND HIT		✓
7	400	14	2	1 ROUND HIT	✓	
8	600 (+/- 100)			1 ROUND HIT	✓	
				TOTALS	6	

16. COMMENTS

17. NUMBER OF ENGAGEMENT MET *(Choose One)*
☐ 8 - EXPERT
☑ 6 - MARKSMAN
☐ 7 - SHARPSHOOTER
☐ 5 AND BELOW - UNQUALIFIED

18. GUNNER'S SIGNATURE: Robert T. Gunning
19. GRADER'S SIGNATURE: Thomas Corley

DA FORM 7521-R, AUG 2003 APD V1.00

Figure 4-30. Example of completed DA FORM 7521-R, MK 19 night practice and qualification with pop-up targets or without NVDs.

CHAPTER 5
TECHNIQUES OF FIRE

Firing techniques allow the gunner to deliver controlled, effective fire on target. This chapter covers the characteristics and classes of fire; range determination; application of fire; firing under degraded conditions; and predetermined fires.

Section I. CHARACTERISTICS AND CLASSES OF FIRE

To effectively employ the MK 19, gunners must understand the definitions of the terms used to describe the actions and effects of firing the MK 19. This section discusses the characteristics and classes of fire.

5-1. CHARACTERISTICS OF FIRE

The following are definitions of the characteristics of fire:

a. **Trajectory.** The path of the projectile in flight. The path is curved due to gravity and elevation of the barrel. The trajectory increases as the sights are elevated for far targets.

b. **Line of Sight.** An imaginary line from the gun to the target, as seen through the sights.

c. **Ordinate.** The vertical distance that occurs anywhere between the line of sight and the trajectory.

d. **Maximum Ordinate.** The highest point of trajectory, at which the vertical distance between the trajectory and line of sight is greatest.

e. **Danger Space.** The area in which the impact of the round or the dispersal of fragmentation causes injuries to personnel or destruction of the target.

f. **Dead Space.** The area(s) in which personnel or targets are safe from the gun's fire. Dead space can be, ditches, depressions, or ravines.

g. **Cone of Fire.** The dispersion of the rounds as they leave the gun.

h. **Beaten Zone.** The area in which the cone of fire strikes the ground or target. Terrain slope and range to the target affect the beaten zone. If the cone of fire falls on upward-sloping ground, the beaten zone is shortened, and vice versa. As range increases, the beaten zone is shortened and widened.

i. **Center of Impact.** The area where the heaviest concentration of fire impacts.

5-2. CLASSES OF FIRE

Fire is classified with respect to the ground, target, and the MK 19.

a. **Respect to the Ground.** Normally, this means either plunging or grazing fire. However, since grazing fire is not practical for use with the MK 19, only plunging fire will be considered. Plunging fire strikes the target from a high angle and confines the danger space to the beaten zone. For example, when fired from the top of a hill, projectiles follow an arcing trajectory and land in the valley (Figure 5-1).

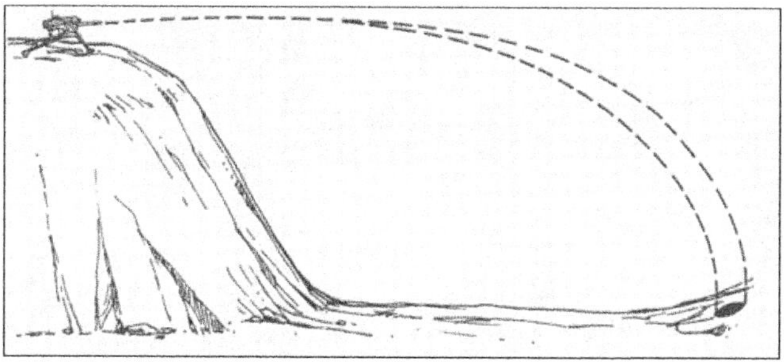

Figure 5-1. Plunging fire.

b. **Respect to the Target.** This class of fire is divided into four types of fire (Figure 5-2a and 5-2b):

(1) ***Frontal.*** The long axis of the beaten zone is at a right angle to the long axis of the target.

(2) ***Flanking.*** This type of fire is delivered against the flank of the target.

(3) ***Oblique.*** This type of fire is directed at a target moving at any angle other than directly toward or perpendicular to the gun.

(4) ***Enfilade.*** This type of fire occurs when the long axis of the beaten zone coincides with the long axis of the target. Enfilade fire may be frontal or flanking, depending on which way the target is facing. For example, frontal enfilade occurs if the MK 19 is in the middle of a road and the target is approaching on the same road. Flanking enfilade occurs if the target is moving either way, left or right, yet is still in a long axis configuration away from the MK 19.

Figure 5-2a. Classes of fire with respect to the target: frontal, flanking, and oblique fires.

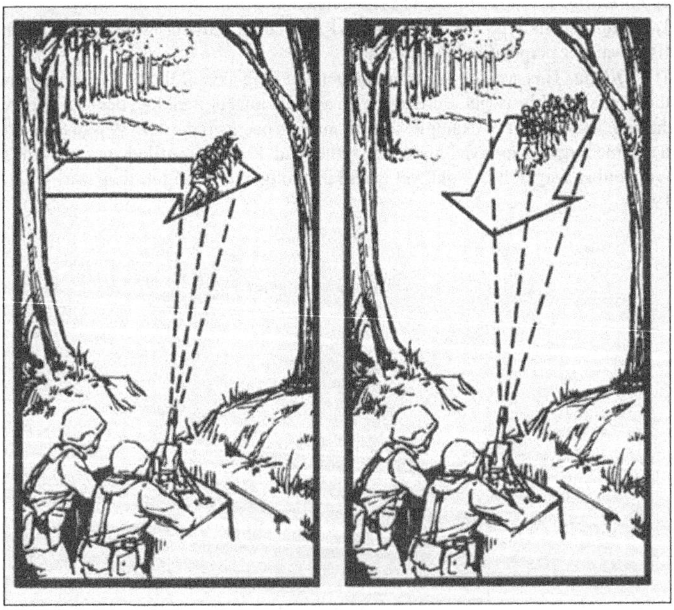

Figure 5-2b. Classes of fire with respect to the target: flanking and frontal enfilade fires.

c. **Respect to the MK 19.** Fire from the MK 19 may be conducted in six ways (Figure 5-3).

(1) ***Fixed.*** Fire is delivered against a target with one aim point, which concentrates the beaten zone.

(2) ***Traverse*** Fire is moved from left to right or right to left, with no range change. It may be used against frontal or flanking targets.

(3) ***Search.*** Fire is directed against a deep target. Elevation changes are made, but direction changes are not. Searching fire is used against enfilade targets.

(4) ***Traverse and Search.*** This is a combination of traversing and searching fires used against a target with depth and width (most likely an oblique target).

(5) ***Swinging Traverse.*** This fire is slightly different from traversing and searching. Although it is delivered against a wide target, with major changes in direction, no elevation changes are made. To deliver fire using a swinging traverse, the gunner releases the traversing slide lock, allowing the gun to travel freely across the traversing bar.

(6) ***Free Gun.*** Fire is delivered with the T&E mechanism removed and is used when quick changes in direction and depth are needed to engage moving targets.

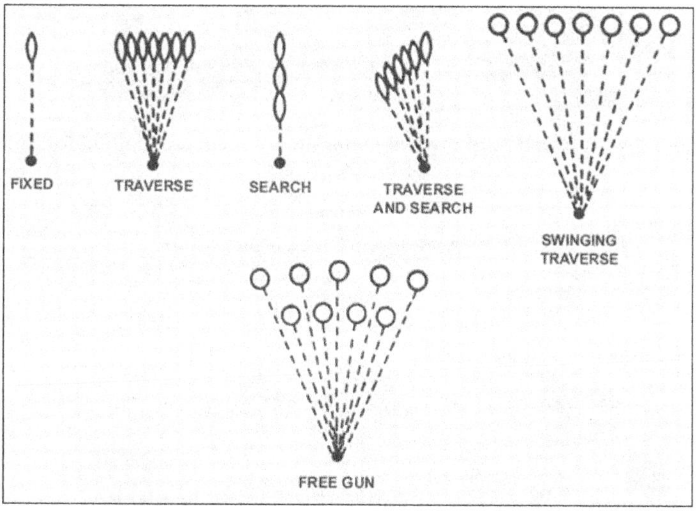

Figure 5-3. Classes of fire with respect to the MK 19.

Section II. RANGE DETERMINATION

Range determination is the process of finding the distance between the gunner's position and his target. The gunner's ability to engage a target effectively depends on determining the correct range to the target.

5-3. MEASUREMENT BY VISUAL ESTIMATION

Range is often determined by this method. There are two ways to estimate range visually.

a. Using the 100-meter unit-of-measure method, visualize 100 meters on the ground (this takes practice) and determine how many units of 100 meters there are to the target (Figure 5-4). For targets more than 500 meters (5 units) away, pick a point about halfway, estimate the range to that point, and double it (Figure 5-5).

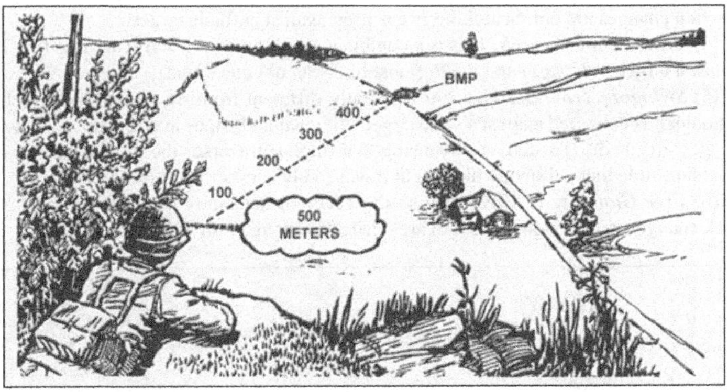

Figure 5-4. Application of the 100-meter unit-of-measure method for ranges up to 500 meters.

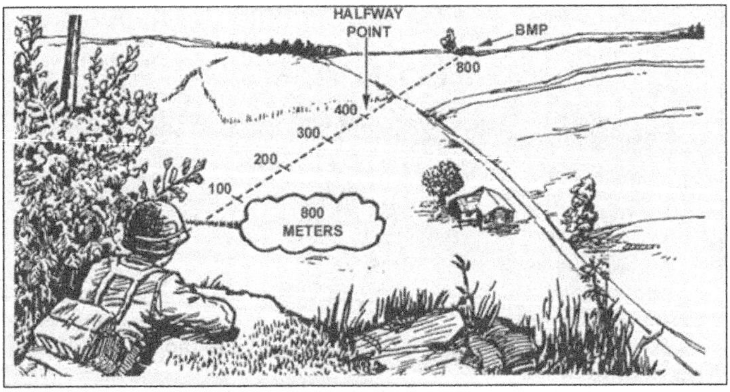

Figure 5-5. Application of the 100-meter unit-of-measure method for ranges greater than 500 meters.

b. Use the appearance-of-objects method, which is fairly reliable, to determine range (this also takes practice). Memorize the sizes and shapes of familiar objects at different ranges. Consider the factors in Table 5-1, which affect the appearance of objects.

FACTORS AFFECTING RANGE ESTIMATION	FACTORS CAUSING UNDERESTIMATION OF RANGE	FACTORS CAUSING OVERESTIMATION OF RANGE
The clarity of outline and details of the target. Nature of terrain or position of the observer. Light and atmosphere.	When most of the target is visible and a clear outline can be seen. When looking across a depression that is mostly hidden from view. When looking downward from high ground. When looking down a straight, open road or along a railroad. When looking over uniform surfaces like water, snow, desert, or grain fields. In bright light or when the sun is shining from behind the observer. When the target is in sharp contrast with the background or is silhouetted because of its size, shape, or color. When seen in the clear air or high altitudes.	When only a small part of the target can be seen or the target is small in relation to its surroundings. When looking across a depression that is totally visible. When looking from low ground toward high ground. When vision is narrowly confined as in streets, draws, or forest trails. In poor light such as dawn and dusk; in rain, snow, fog; or when the sun is in the observer's eyes. When the target blends into the background or terrain.

Table 5-1. Factors affecting visual range estimation.

5-4. MEASUREMENT FROM A MAP

Another way range may be determined is by using a military map. The gunner locates his position and the target's position on the map. He measures the distance and uses the legend scale at the bottom of the map to find the range.

5-5. MEASUREMENT BY PACING

When using this method, the leader ensures there is no immediate danger to the gun's crew. The crew sets up its position, and the gunner selects the target. The assistant gunner walks to the target in a straight line, counting the number of paces it takes to reach the target. As the distance is paced off, the gunner can determine dead space when the assistant gunner disappears from view.

5-6. MEASUREMENT USING BINOCULARS

Binoculars can be used to determine range. The recently adopted M19 binocular, unlike other models, is a lightweight, compact instrument intended for use in general field observation and fire direction. The binocular's left lens includes horizontal and vertical reticles graduated in 10-mil increments (Figure 5-6). When using binoculars to determine range, the soldier must understand the mil relationship (Figure 5-7).

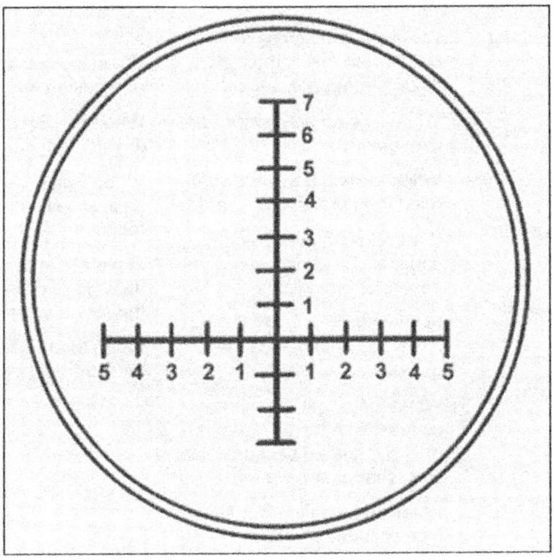

Figure 5-6. M19 binocular reticle.

a. The mil (m̸) is the unit of angular measurement used in adjusting fire. A circle has 6,400 mils. At a distance of 1,000 meters, an object 1 meter wide measures 1 mil. Change mils to meters by multiplying the number of mils times the range (distance) in thousands of meters. Obtain the unknown width or range to an object using a similar method (Figure 5-7).

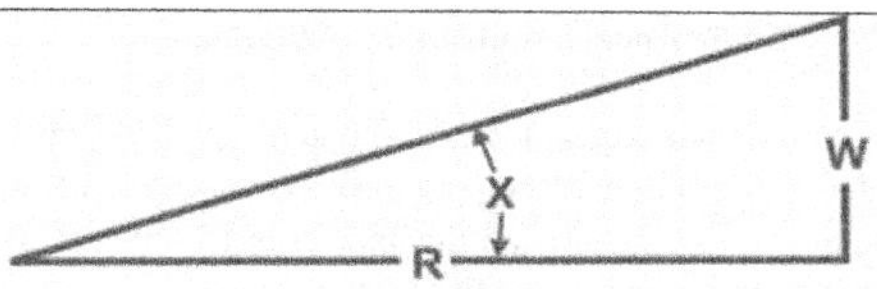

X - IS THE ANGULAR WIDTH OF THE OBJECT IN MILS (m̸).

W - IS THE WIDTH OF THE OBJECT IN METERS.

R - IS THE RANGE OR DISTANCE IN THOUSANDS OF METERS.

FORMULA: $X = \frac{W}{R}$

Figure 5-7. Mil relationship.

b. The mil relationship, as shown in Figure 5-7, is X = W ÷ R; where X is the angular width of the object in mils (m̸), W is the width of the object in meters, and R is the range or distance in thousands of meters.

(1) To find <u>W</u>idth when the known <u>R</u>ange is 4,000 meters and the object is 15 m̸ wide, multiply R times X m̸ or 4 (range in thousands) times 15 (mils). The answer is 60 (meters), or 4 x 15 = 60 meters.

W = R times Xm̸

W = 4 times 15

W = 60

(2) To find <u>R</u>ange when the known <u>W</u>idth in meters (between two bursts or two objects) is 60 and the angular measurement for the same width, when measured with binoculars, is known to be 15 mils, divide (W: m̸) 60 (meters) by 15 (mils).

The answer is 4,000 meters:

$$R = \frac{W}{X\ m\!\!\!/}$$

$$R = \frac{60}{15}$$

R = 4 (in thousands of meters) = 4,000 meters

(3) To find the angle A in m̸ when the known Width in meters between a reference point and the target is 60 meters and the known Range to the target is 4,000 meters, divide W by: R or 60 (meters) by 4 (range in thousands) and the answer is 15 mils (mils):

$$A\ m\!\!\!/ = \frac{W}{R}$$

$$A\ m\!\!\!/ = \frac{60}{4}$$

$$A\ m\!\!\!/ = 15$$

c. The observer uses the mil relation in computing direction shifts as indicated in Figure 5-8.

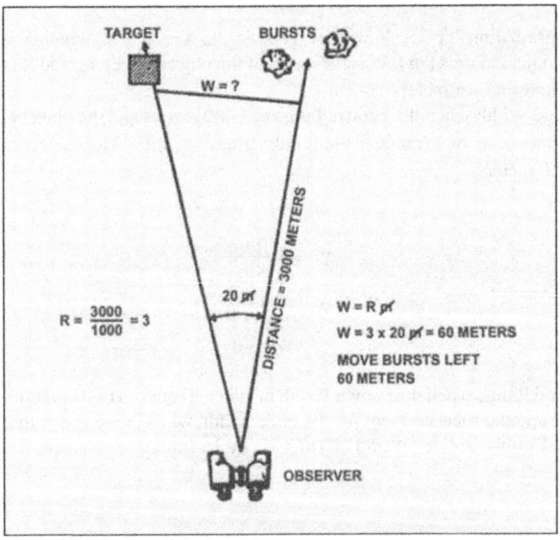

Figure 5-8. Computation of direction shift.

5-7. MEASUREMENT USING AN/GVS-5 LASER RANGE FINDER

The handheld AN/GVS-5 laser range finder provides observers and small-unit leaders the ability to more accurately determine range to targets and other known points. This range finder allows first-round fire for effect, which results in increased enemy casualties and saves ammunition. Friendly units can use the range finder to precisely locate themselves using intersection and modified resection techniques. The range finder aids in the determination of sector depth, range to suspect enemy locations, distance to targets along avenues of approach, and information for overlapping fires. When an AN/GVS-5 is available, it should be used to construct range cards. Information about placing the AN/GVS-5 into action can be found in TM 11-5860-201-10.

5-8. MEASUREMENT BY FIRING

The gunner can determine range by firing a zeroed MK 19. The gunner fires a burst and uses the T & E mechanism to "walk" the rounds on target. The gunner sets and reads the sights on the target, and notes the range. This measurement method is used frequently in combat situations.

5-9. MEASUREMENT OF LATERAL DISTANCE

The gunner can use the finger method to find distance. He extends his arm toward the target with his palm out, elbow locked, one eye closed, and index finger raised (Figure 5-9). He sights along the edge of the finger, adding extra fingers to fill in the space (the average finger is 30 mils wide). One finger equals 30 mils, two fingers equals 70 mils, three fingers equals 100 mils (Figure 5-10).

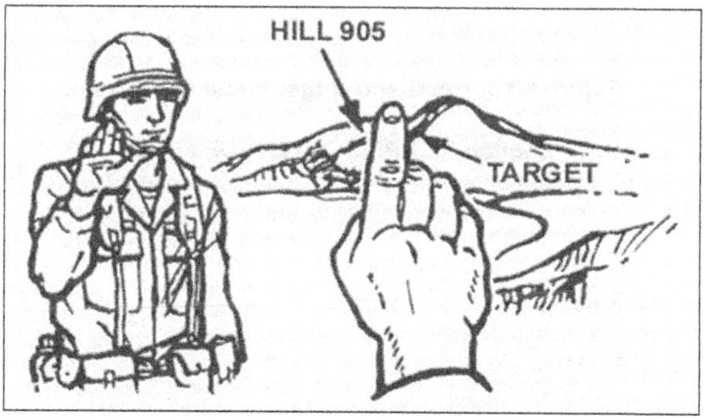

Figure 5-9. Use of fingers to measure deviation.

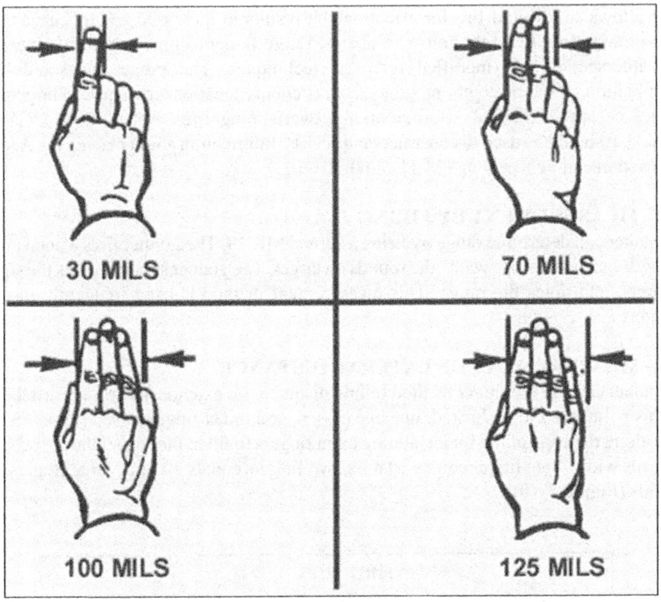

Figure 5-10. Hand-and-finger value estimates.

Section III. APPLICATION OF FIRE

This section includes information about MK 19 fire control and target engagement. Fire control includes all operations connected with the preparation and application of fire to a target.

5-10. FIRE CONTROL

To correctly exercise fire control, the gunner must be able to:

- Open fire at the instant desired.
- Adjust fire on the target.
- Regulate the rate of fire.
- Shift fire from one target to another.
- Cease fire.

Failure to exercise correct fire control results in danger to friendly troops, loss of surprise, premature disclosure of position, misuse of fire on unimportant targets, loss of time in securing adjustments, and waste of ammunition. The main factors to consider for effective fire control are as follows:

a. **Sector of Fire.** This is an area of terrain that an assigned unit covers. It has specific, designated boundaries. Sectors of fire vary in size but are generally limited areas that can be engaged without movement of the tripod.

b. **Rate of Fire.** Surprise and shock effect may be obtained by firing all MK 19s at the same time, using the rapid rate for at least the first few bursts. Engage fleeing targets as soon as possible and with all available fire. Use the rapid rate for the initial delivery of fire to make adjustment of fire easier. Also, whenever the rate of fire is not stated, use the rapid rate. In all cases, unless otherwise ordered, fire the first few bursts at the rapid rate; thereafter, use the prescribed rate.

c. **Adjustment of Fire.** Adjustments may be made before or after firing.

(1) ***Initial Adjustment.*** Set the sights on the range to the target, lay on the target, fire an aimed burst, and observe the impact. When the initial burst is correct, continue to fire (manipulating if necessary) until the target is covered. When the initial burst is not correct, determine from its impact the amount of traverse and search required to place the next burst on the target. Manipulate the gun with the T&E handwheels, making large range corrections by resetting the sights and re-laying on the target.

(2) ***Subsequent Correction and Adjustment.*** Observation and adjustment of fire is the most important element of fire control. It is continuous throughout the action. The gunner is trained to observe and adjust fire without command and to check the lay of the gun frequently. The gunner is also trained to foresee the action of the enemy after he opens fire, and to shift fire to cover any changes in the formation or location of the target. If the gunner fails to do these things, the leader promptly corrects him by announcing or signaling subsequent fire commands. This responsibility to adjust fire continues up through the chain of command. When subsequent fire commands are issued, the gunner makes the required adjustments. When the gun is fired from the tripod mount, subsequent commands are given as adjustments to the elevation or deflection at which the last burst was fired. The adjustments are given in mils and are announced or signaled as ADD, DROP, RIGHT, or LEFT so many mils. Changes in deflection and elevation such as RIGHT 2 or ADD 5 are made with the T&E handwheels.

5-11. FIRE COMMANDS

Fire commands are the means by which leaders control fires. A fire command is a set of instructions, given in a prescribed format that enables the gun crew to properly engage the desired targets. There are two types of fire commands: initial and subsequent. Initial fire commands are used for engagement of targets and the shifting of fire to new targets. Subsequent fire commands are used to adjust fire, change the rate of fire, and cease-fire. The explanation below is based on two gun crews. A good fire command is brief and clear, and in the proper sequence. It is given clearly at a rate that is easily understood by the gunner. Gunners repeat each element to ensure understanding. It is unlikely for a complete initial fire command to be issued during a fire fight. The leader determines which elements of a fire command are obvious to the gunners and which elements must be given to them. Some targets may be engaged with only the alert, range, and command of fire. For example: FIRE MISSION, SEVEN HUNDRED, FIRE. The procedures outlined below are used to accustom gunners to issuing and executing instructions in a logical sequence. The use of complete fire commands in training prepares the gunners to receive fragmentary fire commands in combat.

a. **Initial Fire Command.** This contains the following elements:

- ALERT.
- DIRECTION (as needed).
- RANGE.

- DESCRIPTION (as needed).
- ASSIGNMENT/METHOD (division, manipulation, rate; all only as needed).
- CONTROL.

(1) ***Alert.*** The alert is the first element of the initial fire command. Its purpose is to choose the gun crews and ready them to receive and execute the fire command. FIRE MISSION is announced for all targets. When two or more guns are to fire, the leader announces, FIRE MISSION. If only one gun is to fire, then NUMBER 1 (2) FIRE MISSION is given. When the leader wants to alert all guns, but only wants one gun to fire, the announcement is FIRE MISSION, NUMBER 1 (2).

(2) ***Direction.*** There are several ways to designate target direction:

(a) *Oral.* When the target is not obvious, the leader must tell the gunners where to look. The leader gives direction as: FRONT, RIGHT FRONT, LEFT FLANK, and so on. The leader may identify an indistinct target by the use of a reference point. The selected reference point, announced as REFERENCE, must be a distinctive terrain feature or object in or near the target area. The word TARGET precedes the target description when a reference point is used.

- When the reference point is within the target area, the leader may describe the target as extending so many mils, meters, or fingers from that point. He announces range to the reference point. For example:
  REFERENCE: BUNKER, FIVE ZERO ZERO.
  TARGET: TROOPS EXTENDING RIGHT TWO FIVE, LEFT TWO FIVE.
- If the selected reference point is outside the target area, the leader announces the range to the target. For example:
  REFERENCE: LONE TREE.
  TARGET: MACHINE GUN IN EDGE OF WOODS; FIVE ZERO ZERO.
- The leader may identify the direction to a target that is not obvious by selecting an obvious feature, and (by naming successive reference points) leading the gunner to the target step by step. For example:
  REFERENCE: RED-ROOFED HOUSE, RIGHT OF HOUSE, HEDGE, CENTER OF HEDGE, GATE, ABOVE GATE.
  TARGET: MACHINE GUN.
- For a tripod-mounted MK 19, the interval between the reference and the target is measured by laying the gun on the reference point and manipulating the gun a given number of mils to the target.

(b) *Firing a MK 19.* Pointing out an obscure target by firing a MK 19 is simple, fast, and accurate. However, it may cause loss of surprise or premature disclosure of the gun's position. The leader announces the general direction of fire, if it is not obvious. He then lays one gun on the target, commands: WATCH MY BURSTS, and fires one or more on the target. The leader completes the designation orally using the target's midpoint or extremes: RIGHT FLANK, MIDPOINT, or NEAR END. Firing rifle tracers at the target also makes it easier for the gunners to find.

(c) *Laying a MK 19.* Laying the gun on a target is a simple and accurate method that does not sacrifice surprise. The leader goes to each gun, lays it on the target, and has the gunners check the lay. The gunners open fire at the same time on command.

(3) ***Range.*** This element follows the target description. The leader announces it in even digits, hundreds or thousands. For example: FOUR FIVE ZERO; THREE HUNDRED; ONE ONE HUNDRED; ONE THOUSAND.

(4) ***Description.*** The leader gives a brief target description to inform the gunners of the nature of the target. Examples of target descriptions include TROOPS, which refers to any dismounted enemy personnel; MACHINE GUN, which refers to any automatic gun; and TANK, which refers to any armored vehicle. If several targets are in view, the leader may describe the target or part of a target to be engaged as LEADING TRUCK, RIGHT BUILDING, FAR END, HALTED COLUMN, and so on. If the target is obvious, no description is necessary.

(5) ***Assignment Method.*** This element is used only when specific assignments are required to divide or subdivide the target, assign class of fire, or announce a rate of fire.

(a) *Division (or subdivision).* The leader gives this element only when it is required (paragraph 5-12). Division may be announced as follows:

NUMBER 1 RIGHT HALF, NUMBER 2 LEFT HALF.
NUMBER 1 RIGHT TWO-THIRDS, NUMBER 2 LEFT TWO-THIRDS.
NUMBER 1 RIGHT TWO-THIRDS, NUMBER 2 LEFT ONE-THIRD.

(b) *Manipulation.* The leader gives this element to prescribe the class of fire with respect to the gun, and gives it only if the required manipulation is not obvious. For example:

FIXED.
TRAVERSE.
SEARCH.
TRAVERSE AND SEARCH.
SWINGING TRAVERSE.

(c) *Rate.* The greatest effect results from having two guns open fire at the rapid rate, at the same time. Regardless of the rate announced, gunners open and adjust fire at the rapid rate, and use the announced rate (sustained, rapid, or cyclic) thereafter. Size of target, type of target, and ammunition supply are the factors that influence selection of a rate of fire. Rates of fire are distinguished as follows:

- **Sustained.** The sustained rate of fire is measured in rounds for each minute and is the rate at which a MK 19 may be fired indefinitely without damage from overheating (40 rounds per minute). The leader announces SUSTAINED.
- **Rapid.** The rapid rate is measured in rounds for each minute and is the rate at which a MK 19 may be fired for a limited time without danger of overheating (60 rounds per minute). This rate permits a high volume of fire to be delivered for a short, set time. Gunners use the rapid rate unless another rate is announced.
- **Cyclic.** The cyclic rate is the rate at which a MK 19 can be fired for a limited time only, at the fastest rate the gun will fire (325 to 375 rounds per minute). This rate may be used when the target or area needs the greatest possible suppression. The leader announces CYCLIC.

(6) ***Control.*** The leader uses this element to give the command to open fire. For immediate engagement of the target, the leader gives the command, FIRE, or the arm-and-hand signal to fire without pause. Fire is often withheld for surprise and effect, and both guns open fire at the same time. To ensure this, the leader may preface the command or signal to fire with the words AT MY COMMAND or ON MY SIGNAL. When the gunners are ready

to engage the target, they report UP or announce: NUMBER 1 (2) UP, READY, and the leader gives the command or signal to fire.

b. **Subsequent Fire Commands.** The leader uses these to repeat or correct a fire command, to adjust fire, to cease or commence fire, or to terminate the alert. The deflection correction must always be given first.

(1) ***Repeat Fire Commands.*** If the gunner fails to understand any element of the fire command, a repetition of the element may be requested by announcing the misunderstood element with rising inflection to denote a question. When repeating any portion of the fire command, the leader will preface it with the words THE COMMAND WAS.

(2) ***Correct Fire Commands.*** The leader changes or corrects a portion of the fire command by announcing CORRECTION and giving a command. For example, to change an incorrect range command of 500 meters to 600 meters, the command is CORRECTION, SIX HUNDRED.

(3) ***Adjust Fire.*** The leader adjusts fire if the initial burst was not on target. The leader does this by announcing the direction and elevation needed to get on target ADD 100, RIGHT 20; or DROP 200, LEFT 50.

(4) ***Cease and Commence Fire.*** CEASE FIRE is announced if the leader wishes to interrupt for any reason. This type of subsequent fire command informs the MK 19 crew that it will remain on the alert and that more instructions will follow. Firing is renewed by announcing a subsequent fire command or by announcing a new fire command. Firing is resumed with the same data by using the command FIRE.

(5) ***Terminate the Alert.*** To allow the MK 19 crew to relax between fire missions, termination of the alert is announced as CEASE FIRING, END OF MISSION.

5-12. TARGET ENGAGEMENT

The method chosen depends on terrain, target presentation, type of target, and tactical situation.

a. **Distribution of Fire.** To be effective, fire must be distributed over an entire target. Improper distribution results in gaps between beaten zones and allows some of the enemy to escape or to use guns without effective opposition.

(1) ***Factors Affecting Distribution of Fire.*** No fixed rule can be given as to the widest target that a single MK 19 may effectively engage. Ideally, the target should be no more than 100 mils wide. The traversing screw on the T&E mechanism limits the gun's traverse unless the traversing slide lock lever is unlocked. Wider targets require more traversing time, which prevents the gunner from placing a continuous volume of fire on the whole target. The engagement of a wide target by a single MK 19 requires excessive ammunition.

(2) ***Fire Unit.*** This consists of a pair of MK 19s. If possible, at least two guns should be assigned to the same mission, although sometimes a single MK 19 may be effective. The assignment of a pair to a single mission ensures continuous fire in case either gun is put out of action. Two guns can provide a great volume of fire on the target, and can reduce the time required to cover it.

(3) ***Manipulation of the Tripod-mounted MK 19.*** Traversing fire is moved in 5-mil increments. The MK 19s are fired after each manipulation to ensure the beaten zones overlap. Searching fire is often used on level or evenly sloping ground. When the ground is irregular, however, the amount of search to apply between bursts in order to ensure overlap of the beaten zones is determined by observation.

b. **Engagement of Point Target.** Any target no larger than the beaten zone is a point target, and is engaged by fixed fire. The command is FIXED. MK 19 crews are trained to follow any movement or change in formation made by the enemy after the initial burst of fire. An example of a fire command for a point target is as follows:

FIRE MISSION.
FRONT.
MACHINE GUN.
SIX HUNDRED.
FIXED.
RAPID.
FIRE.

c. **Engagement of Linear Target.** There are several ways to engage a linear target. The method used is chosen based on the number of MK 19s available and whether or not the entire target is visible to the gunner.

(1) ***Linear Engagement with One MK 19.*** A single MK 19 engages the target the same as either one of a pair. The MK 19 is laid just outside either flank (or on a reference point within the target area) and the gunner adjusts fire on the flank (or point). The gunner traverses back and forth across the entire area until told to cease firing. The leader may designate where he wants initial fire. For example, if he directs a single gunner to engage a target with width he may announce:

FIRE MISSION.
FRONT.
TROOPS EXTENDING FROM DEAD TREE RIGHT 20 MILS.
FIVE HUNDRED.
TRAVERSE.
RAPID.
AT MY COMMAND.
FIRE (given after the gunners announce "Up").

(2) ***Linear Engagement with Two MK 19s.*** A linear target may be engaged two ways when a pair of MK 19s is used, depending on the width of the target in mils.

(a) *Target 100 mils or Less in Width.* The normal traversing method is used in this case. Each MK 19 delivers the initial burst of fire on its corresponding flank of the target. Fire is adjusted on that point. Each MK 19 is traversed across the target to the other flank, covering the entire target, until the gunners are told to cease firing (Figure 5-11). The command for this type of fire is TRAVERSE. An example of a fire command used for this situation follows:

FIRE MISSION.
RIGHT FRONT.
TROOPS, EXTENDING FROM DEAD TREE RIGHT TO CLEARING.
SEVEN FIVE ZERO.
TRAVERSE.
RAPID.
AT MY SIGNAL.
FIRE (given after the gunner announces "Up").

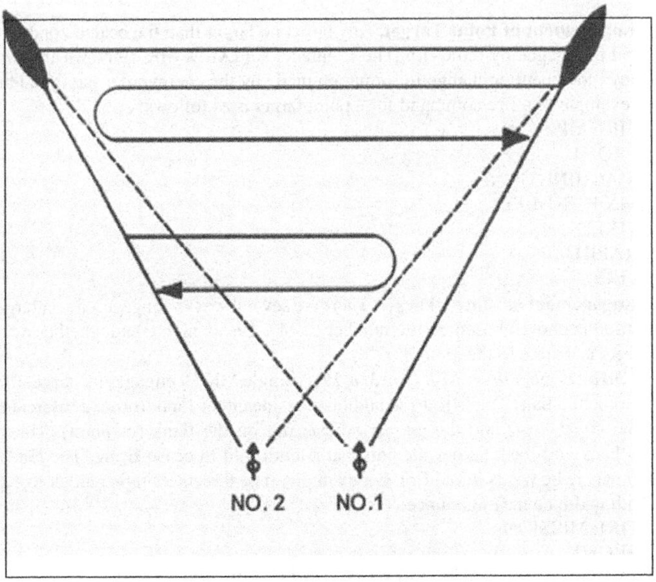

Figure 5-11. Traversing method, pair of MK 19s, target 100 mils or less in width, both flanks visible.

(b) *Target More than 100 mils in Width.* In this case, the leader assigns part of the target to each MK 19 (Figure 5-12). One part may be smaller than the other to ensure it receives a heavier concentration of fire. Each gun fires an initial burst on its respective flank, and covers its assigned part of the target as described in subparagraph (a). An example of a fire command follows:

FIRE MISSION.
RIGHT FRONT.
TROOPS, EXTENDING FROM DEAD TREE RIGHT TO CLEARING, ROW.
EIGHT HUNDRED.
NUMBER 1, RIGHT ONE-THIRD; NUMBER 2, LEFT TWO-THIRDS.
TRAVERSE.
RAPID.
AT MY COMMAND.
FIRE (given after the gunners announce "Up").

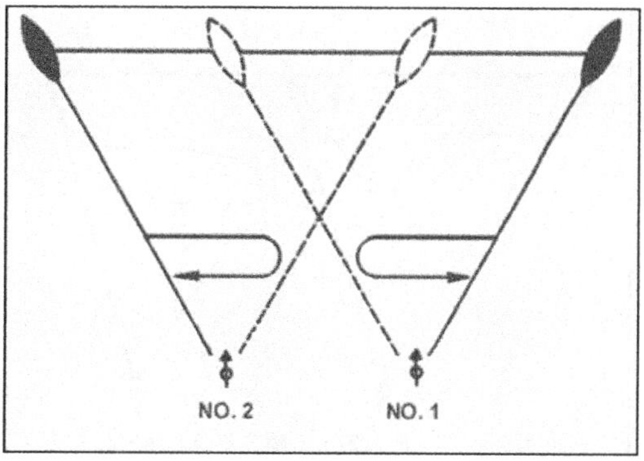

Figure 5-12. Traversing method, pair of MK 19s, target more than 100 mils in width.

(3) ***Linear Engagement with Four MK 19s.*** Four guns may be needed to concentrate a heavy volume of fire into a wide target area. Methods used are similar to the ones described for a pair of MK 19s. The width of the target in mils determines which type of linear engagement is used.

(a) *Target 100 to 200 mils in width*:

- Each pair of MK 19s engages the whole target (Figure 5-13).

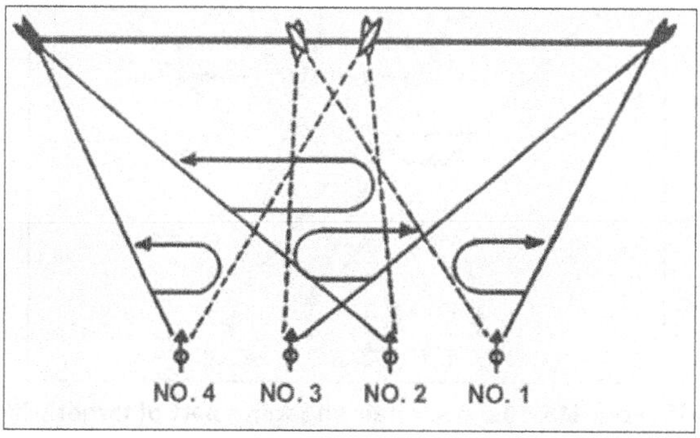

Figure 5-13. Four MK 19s, target 100 to 200 mils in width.

- Each pair of MK 19s engages half the target (Figure 5-14).

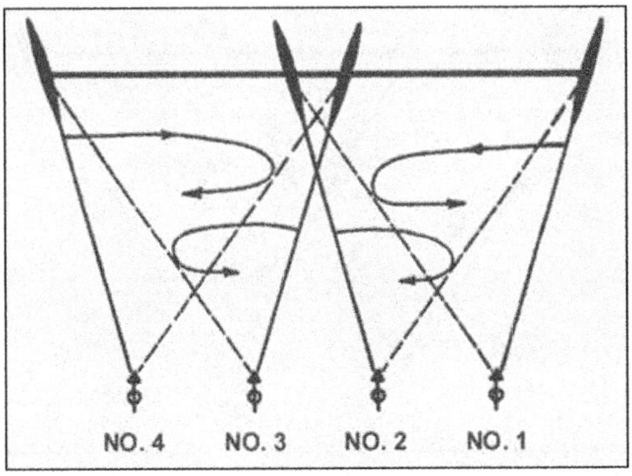

Figure 5-14. Two pairs of MK 19s engaging each half of a target 100 to 200 mils in width.

(b) *Target Greater than 200 mils in Width.* The leader may assign part of the target to each gun in this case. The target may be divided in half for each fire unit or in four parts, one for each MK 19 (Figure 5-15).

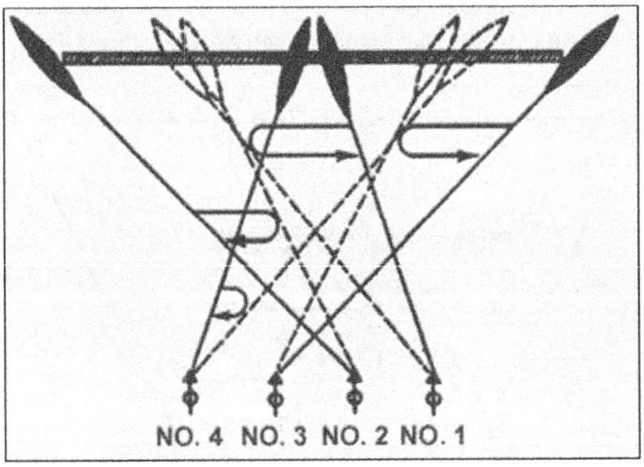

Figure 5-15. Four MK 19s, each pair engaging part of target with extreme width.

(4) ***Linear Engagement with the Flanks of Target Not Identifiable.*** The target may be located by firing tracers from an observer's gun, laying MK 19s, or using a reference point. The reference point may be in or near the target area.

(a) When a reference point is outside the target area, the leader may direct the gunner to it by announcing the interval (right or left, long or short) between the reference point and the target. With the MK 19 on a tripod mount, the gunner measures the interval right or left between the reference point and the flank of the target by laying on the reference point, with the sights set at the range to the target, and then shifting the MK 19 the designated number of mils. The interval long or short may be measured in mils using the computed search method, or it may be estimated in meters. When the gunners lay each MK 19 on its respective flank, they adjust fire and engage the target.

(b) When a reference point is in the target area, the leader may identify the flanks of an obscure target as extending so many mils from the reference point. In this case, gunners use the swinging traverse. With the swinging traverse each MK 19 is laid on the announced reference point (initial aiming point), adjusted for fire, and traversed the given distance to its corresponding flank and back, firing after each manipulation (Figure 5-16). Each gunner continues traversing back and forth across the entire target until told to cease-fire. Each gunner stops firing while traversing past the reference point. Examples of fire commands that may be used are as follows:

- The leader designates the target by firing one gun:

  FIRE MISSION.
  LEFT FLANK.
  WATCH MY BURSTS (OR TRACERS).
  (Lays and fires MK 19 at left flank) LEFT FLANK.
  (Lays and fires MK 19 at right flank) RIGHT FLANK.
  TROOPS.
  NINE HUNDRED.
  TRAVERSE.
  RAPID.
  AT MY COMMAND.
  FIRE (given after gunners announce "Up").

- The leader designates the target by using a reference point outside the target area (MK 19s on tripod mounts):

  FIRE MISSION.
  RIGHT FRONT.
  REFERENCE: DEAD TREE.
  RIGHT SIX ZERO MILS; DROP THREE FIVE ZERO METERS
  (or drop so many mils, if using computed search).
  TARGET: TROOPS EXTENDING RIGHT FIVE ZERO MILS.
  SIX HUNDRED.
  TRAVERSE.
  RAPID.
  AT MY COMMAND.
  FIRE (given after the gunners announce "Up").

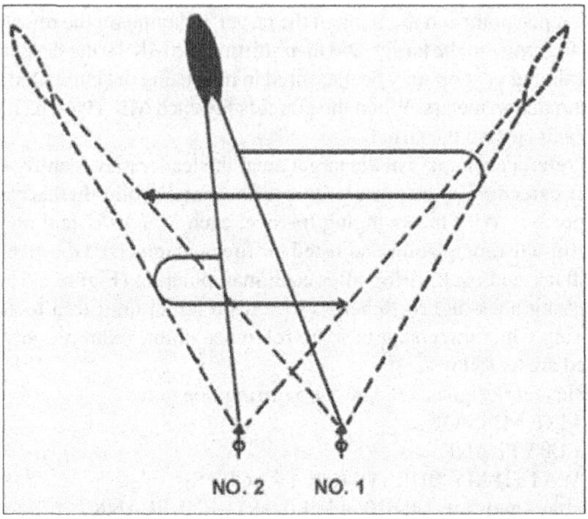

Figure 5-16. Two MK 19s, deep enfilade target, ends not visible.

- The leader designates the target by using a reference point within the target area (MK 19s on tripod mounts):

  FIRE MISSION.
  FRONT.
  REFERENCE: LONE TREE.
  TARGET: TROOPS EXTENDING RIGHT TWO ZERO MILS, LEFT, THREE ZERO MILS.
  SEVEN HUNDRED.
  TRAVERSE.
  RAPID.
  AT MY COMMAND.
  FIRE (given after the gunners announce "Up).

(5) ***Linear Engagement with Swinging Traverse.*** As previously described, a single gun uses this method against massed or rapidly moving targets at short ranges. The leader's fire command for a swinging traverse can be:

FIRE MISSION.
LEFT FRONT.
TRUCKS.
SWINGING TRAVERSE.
FIRE.

d. **Engagement of Deep Targets Using Searching Fire.** There are several ways for leaders to use two guns to engage a stationary deep target, depending on whether the ends of the target are visible to the gunner. If the target location is unknown, the computed search method may be used to engage the target. A deep target that is stationary or that has limited mobility may be engaged with searching fire.

(1) ***Target Ends Visible to Gunner.*** The Number 1 gun is adjusted first on the near end and the Number 2 gun on the far end. Each gunner searches to the opposite end and back again repeatedly, until told to cease firing (Figure 5-17). Target depth is considered in determining sight settings. The command is SEARCH.

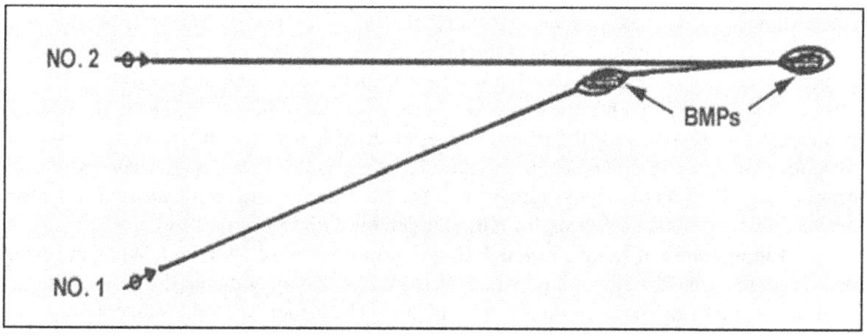

Figure 5-17. Two MK 19s, deep enfilade target, ends visible.

(a) *Target Less than 200 meters Deep.* The leader announces the range to the midpoint of the target for both MK 19s, using the length of the beaten zones to ensure the initial bursts impact on the target. An example of a fire command for this situation follows:

FIRE MISSION.
FRONT.
HALTED COLUMN.
SEVEN HUNDRED (midrange).
SEARCH.
RAPID.
AT MY COMMAND.
FIRE (given after gunners announce "Up").

(b) *Target Depth Greater than 200 meters.* The leader announces the range to the near end for the Number 1 gun, and to the far end for the Number 2 gun. An example of a fire command for this situation follows:

FIRE MISSION.
LEFT FRONT.
TROOPS ALONG RIGHT EDGE OF WOODS.
NUMBER 1, SIX HUNDRED; NUMBER 2. NINE HUNDRED.
SEARCH.
RAPID.
AT MY COMMAND.
FIRE (given after gunners announce "Up").

(2) ***Target Ends Not Visible to Gunner.*** The target may be located by having an observer fire tracers at the target, by laying the MK 19s, or by using a reference point. The reference point must be visible to the gunner and in the vicinity of the target. When a selected reference point is outside the target area, the leader may direct the gunner to the target and announce the interval to the right or left (long or short) between the reference

point and the target. If the MK 19 is on a tripod, the gunner lays on the reference point with the sights set at the range to the midpoint of the target; he measures the interval and shifts the appropriate number of mils. Using the computed search method, the gunner may measure the long or short interval in mils, or he may estimate the interval in meters. When the gunners lay their MK 19s on the right or left flank (near or far end), respectively, they adjust fire and engage the target. When the gunners lay their MK 19s on the announced release point (initial aiming point), they set their sights at the range to the reference point, and adjust fire. The gunner on the Number 1 gun searches down the designated number of meters (or mils), firing after each manipulation until the gun is set on the near end of the target. The gunner on the Number 2 gun searches up the designated number of meters (or mils), firing after each manipulation, until the gun is set on the far end of the target. When each gun has reached the near or far end, respectively, the gunners reverse the direction and both search up and down between the two limits, covering the entire target, until told to cease firing.

e. **Engagement of Deep Targets Using Computed Search Method.** When the depth of a target must be described in mils (search), the leader may compute the depth from the reference point using the computed search method. The leader determines the ranges to the near and far ends of the target and finds the required AE (angles of elevation) in mils for both ranges (Appendix F, Firing Table). The leader then computes the difference between the two. This is the amount of search required when the MK 19 and target are at the same elevation on level ground. If the fire is plunging, the amount of search should be increased. For example, when using M430 ammunition, a deep target has been sighted (the ends of which are not visible to the gunners). A reference point is also visible on the target at a range of 1,100 meters, the depth of the target is estimated at 200 meters, and the reference point appears to be midway between the ends of the target:

$\frac{200 \text{ meters}}{2}$ = 100 meters from the reference point to each end of target

1,100 + 100 = 1,200 meters to far end of target

1,100 - 100 = 1,000 meters to near end of target

AE 1,200 meters = 175 mils

<u>-AE 1,000 meters</u> = <u>132 mils</u>

depth of target 43 mils

$\frac{43 \text{ mils}}{2}$ = 21.5 mils from release point to each end of target

To search in 5-mil increments, the gunners cover the target by searching 25 mils long and 25 mils short of the reference point, or a total of 50 mils. The leader describes the target to the gunner as TARGET; TROOPS EXTENDING LONG 25 MILS; SHORT 25 MILS. The following are examples of fire commands:

- Reference point in target area, ends of target not visible to the gunner.

  FIRE MISSION.
  RIGHT FRONT.
  REFERENCE: LONE BUSH.
  TARGET: TROOPS EXTENDING SHORT ONE HUNDRED YARDS, LONG TWO HUNDRED YARDS.
  SIX HUNDRED (range to reference point).
  SEARCH.
  RAPID.

AT MY COMMAND.
FIRE (given after gunners announce "Up").

- Reference point outside target area, gun on tripod.

FIRE MISSION.
LEFT FRONT.
REFERENCE: DEAD TREE.
RIGHT THREE ZERO MILS; ADD TWO HUNDRED METERS
(or add so many mils, using computed search).
TARGET: TROOPS EXTENDING OVER TWO HUNDRED METERS
(or long so many mils, using computed search).
NINE HUNDRED.
SEARCH.
RAPID.
AT MY COMMAND.
FIRE (given after gunners announce "Up").

f. **Engagement of Moving Targets.** Due to the slow flight of the MK 19 round, it is not practical to engage moving targets using the track-and-lead methods of engaging moving targets. In order to effectively engage moving targets with the MK 19, use the trapping method of engaging targets. The gunner chooses an aiming point forward of the target and on the target path, and presses the trigger before the target comes into the sights. After starting to fire, the gunner moves the MK 19 slowly toward the target. The target moves into the impact area, and is "trapped." The gunner must begin firing before the target is in the sights or in the impact area (Figure 5-18).

(1) ***Engagement of Deep Target Moving Away from the MK 19 Position.*** If the target is moving rapidly away, both gunners lay their MK 19s on the far end of the target with the range set to that point, and search up.

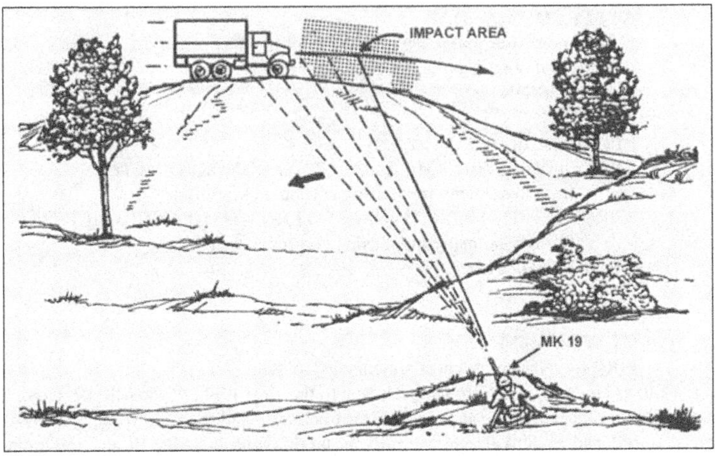

Figure 5-18. Target-trapping method.

(2) ***Engagement of Deep Target Moving toward the MK 19 Position.*** If the target is moving rapidly toward the MK 19 position:

(a) *One MK 19*. The gunner lays the MK 19, adjusts on the near end of the target with a range to the target's midpoint, and searches up and down the entire target. When a reference point within the target area is announced, the gunner lays on it with the range to that point. The gunner adjusts fire and searches up and down the entire target, beginning with the near end, until told to cease firing.

(b) *Two MK 19s*. The gunners lay both guns on the near end of the target, set the sights on the range to that point, and search down. The manipulation element of the fire command for a rapidly approaching or receding target is BOTH WEAPONS; FAR (NEAR) END; SEARCH.

(c) *Four MK 19s*. Each pair of gunners engages the target as if they were acting alone. If one pair must be switched to a different target, the other pair continues to cover the original target. A deep target should never be subdivided since the elevating mechanism on the M3 tripod allows enough search to cover any deep target within the maximum effective range for direct lay (1,500 meters).

g. **Engagement of Linear Target with Depth Using Traversing and Searching Fire.** The method used depends on the number of MK 19s available and whether the flanks of the target are visible to the gunner.

(1) ***One MK 19, Flanks Visible.*** The gunner lays initially on the near flank of the target with range set to the midpoint, adjusts fire on the near flank, and traverses and searches back and forth, covering the entire target, until told to cease-fire.

(2) ***Two MK 19s, Flanks Visible.*** Gunners use the same method as for engagement of a deep linear target: they must search to keep the center of impact on the target. The leader determines the range for the initial fire command the same as for a deep target (Figure 5-19).

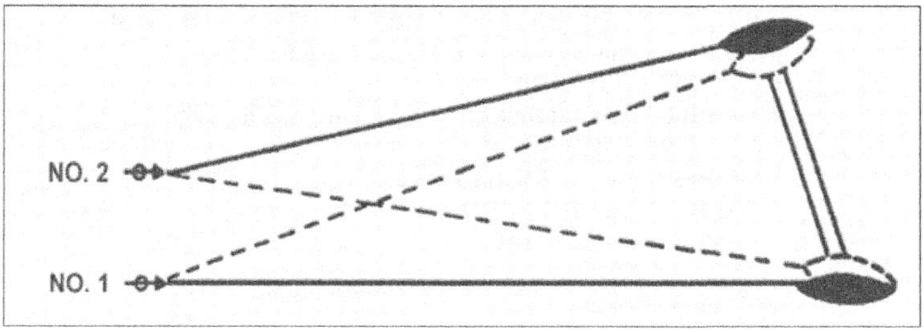

Figure 5-19. Two MK 19s, linear target with depth.

(3) ***Four MK 19s, Flanks Visible.*** Gunners use the same method for two pairs as is used for two guns. If one pair must be switched to another target, the other continues to cover the original target.

(4) ***Flanks Invisible.*** The leader may designate the target using a reference point. However, a reference point in the target area is not used as an initial aiming pointbecause of the difficulty in describing an oblique target.

(5) ***Fire Commands.*** The leader may use the following fire commands for linear targets with depth.

- When the target is not wide enough to require subdivision, is 200 meters or less in depth, and is to be engaged by either one or two MK 19s:
  FIRE MISSION.
  RIGHT FRONT.
  TROOPS EXTENDING FROM BLACK STUMP RIGHT TO LONE PINE.
  SEVEN HUNDRED (midrange).
  TRAVERSE AND SEARCH.
  RAPID.
  AT MY SIGNAL.
  FIRE (given after gunners announce "Up").
- When the target is 100 mils wide or less, is 200 meters deep or less, and is to be engaged by four MK 19s:
  FIRE UNIT, FIRE MISSION.
  FRONT.
  TROOPS TO THE RIGHT OF RED BANK.
  SEVEN HUNDRED (midrange).
  TRAVERSE AND SEARCH.
  RAPID.
  AT MY SIGNAL.
  FIRE (given after gunners announce, "Up").
- When the target is wide enough to require subdividing, is more than 200 meters in depth, and is to be engaged by either two or four MK 19s:
  FIRE MISSION.
  FRONT.

TROOPS EXTENDING FROM TRUCK, RIGHT TO BRIDGE.
NUMBER 1 (and 3 if four MK 19s are engaging the target),
SEVEN HUNDRED.
NUMBER 2 (and 4 if four MK 19s are engaging the target),
ONE THOUSAND.
NUMBER 1 (and 3), RIGHT HALF.
NUMBER 2 (and 4), LEFT HALF.
TRAVERSE AND SEARCH.
RAPID.
AT MY SIGNAL.
FIRE (given after gunners announce "Up").

h. **Engagement of Target Using Combined Sights.** Leaders may choose this expedient way to engage a linear target with depth. They may lengthen the beaten zone of a two- to four-gun fire unit by having gunners set the sights differently on each gun. The leader gives one gunner a setting 150 meters less than the estimated range, and the other a setting 50 meters more than the estimated range. When four guns are used, one pair of gunners sets sights 150 meters less than the range, the other 150 meters more than the range. This technique sacrifices precision for speed, but may be necessary for fast-moving or fleeing targets. A sample fire command for the use of combined sights follows:

FIRE MISSION.
RIGHT FRONT.
TROOPS EXTENDING FROM CROSSROADS TO HOUSE.
1,300 METERS (midrange).
COMBINED SIGHTS; NUMBER 1, 1,350 METERS; NUMBER 2,1,250
METERS.
TRAVERSE.
RAPID.
FIRE (given after gunners announce "Up").

i. **Engagement of Area Target.** This applies to a target that cannot be covered by either traversing fire, searching fire, or traversing and searching fire. Area target engagement requires that the leader mass fires from four to six guns. Examples of area targets include:

- Large troop or mechanized formations.
- Targets that must be suppressed (if the exact positions are unknown or are not visible).
- Large kill zones or engagement areas for which predetermined fires have been planned.

j. **Engagement of Target Using Massed Fires.** Leaders can mass fires in both offensive and defensive operations, depending on the time available to plan and conduct the engagement.

(1) ***Defensive Operations.*** There are two types of defensive operations.

- **Defense, Time Not a Factor.** Massing fires into an engagement area has advantages over other types of area fire engagements. Once TRPs are designated, gunners may fire upon them to register or rehearse the MK 19s. After the target has been subdivided and TRPs registered, the gunners traverse, search up, traverse back, and search down through their assigned parts of the target to assure full coverage of the target (Figure 5-20). The same method applies regardless of how many MK 19s are used. If four or six guns are used, the leader must subdivide the target and assign more TRPs than for a pair of MK 19s. An example of a fire command for this situation follows:

  FIRE MISSION.
  FRONT TRPs 1, 2, 3, 4.
  (MK 19s are laid on respective TRPs)
  AREA: ENEMY COMPANY ENTERING EA GOLD.
  (Range omitted MK 19s on TRPs.)
  TRAVERSE.
  SEARCH UP 50 MILS.
  RAPID.
  AT MY COMMAND.
  FIRE (given after the enemy is completely within the engagement area and the gunners announce "Up").

- **Defense, Time a Factor.** Leaders may not have the luxury of registering their MK 19s on TRPs before chance contacts due to time factors or the elements of surprise. However, the lethal effect of massed MK 19 fire on an area target should not be overlooked. Units should develop and practice SOPs that enable them to engage large-area targets on quick notice. For example, the leader may give each MK 19 a section or quadrant for all quick area engagements. He may assign the upper left side of the target to the Number 1 gun, the upper right to Number 2, and so on.

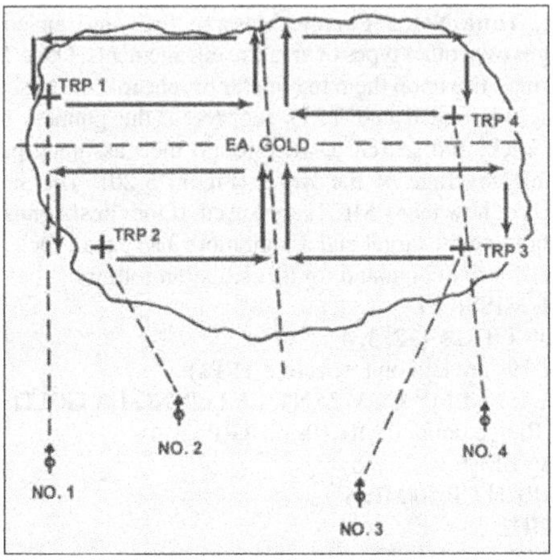

Figure 5-20. Use of massed fires, time not a factor.

(2) ***Offensive Operations.*** Massing fires into an area target can also be useful for offensive operations. During a movement to contact or an attack where time is a factor, massed fires may be used to suppress a suspected or actual enemy position, allowing the unit to flank or bypass the enemy. Although fast target acquisition and volume of fire are primary goals, the leader should subdivide the target for control, and adjust fires for greatest effect. Massed fires are especially useful in support of an attack where time is not a factor, which may require the use of overhead fire.

WARNING

Overhead fire should not be used unless the vertical interval between friendly soldiers and the target line is great enough to ensure safety.

5-13. DEFILADE POSITION

The MK 19 is most effective when laid directly on a target. Depending on the tactical situation, however, defilade could be the most desirable position.

a. **Definition.** A MK 19 is in defilade position when the gun and its crew are hidden by a landmass such as the crest of a hill from enemy ground (Figure 5-21). The position can be on the reverse slope of a hill, on the forward slope of the next high ground to the rear of a hill, or in a small fold in the ground. A defilade firing position does not necessarily reduce the effectiveness of fire against a stationary enemy target or preclude fire over the heads of friendly troops. Defilade may also be useful in providing predetermined fire into an EA.

b. **Advantages.** Some advantages of using the defilade position are that the MK 19 and a crew have cover and concealment from direct fire guns; the crew has some freedom of movement near the MK 19 position; control and supply are facilitated; and the characteristic smoke and flash of the MK 19 are partly concealed from observation.

c. **Disadvantages.** Disadvantages of using the defilade position are that targets close to the mask usually cannot be engaged, and (because adjustment of fire must be made through an observer) rapidly moving ground targets are not easily engaged.

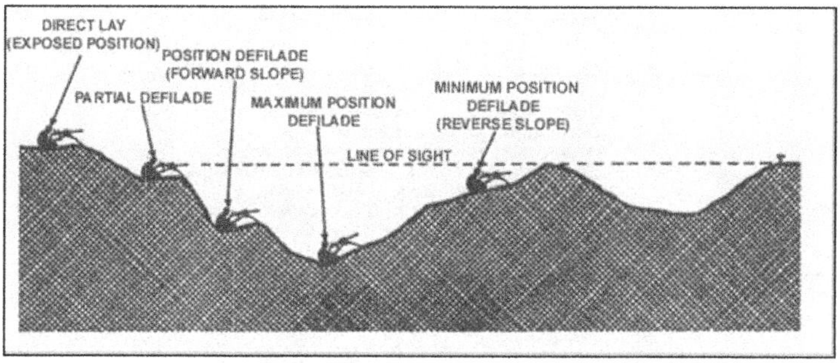

Figure 5-21. Defilade positions (and direct lay).

d. **Options.** A defilade position can either be on a forward or reverse slope.

(1) ***Reverse Slope Positions.*** These positions afford some protection from high-angle enemy fire and direct-fire cannon.

(a) *Partial Defilade Position.* The MK 19 is in partial defilade position when it is sited at the highest point on a slope on which it is defiladed. Partial defilade position provides great flexibility in engagement of new targets, although it does not provide the best cover. The MK 19 crew positioned in partial defilade can be grazed by enemy machine gun fire, but is concealed from enemy observation near the target area.

(b) *Maximum Defilade Position.* The MK 19 is in maximum defilade position when it is sited at the lowest point on a slope from which it can engage a target. The MK 19 crew positioned in maximum defilade has relatively good cover but lacks the flexibility to engage new targets.

(2) ***Forward Slope Defilade Position.*** A forward slope defilade position is not vulnerable to grazing fire from the target area, but it is open to attack or damage from mortar and artillery fire. The MK 19 is in partial defilade on a forward slope if the gun and gunner have some protection from direct fire, and if the gunner is able to engage the target using direct lay. Partial defilade is used only when a fire mission cannot be completed from defilade. The MK 19 is held in defilade and moved into partial defilade to fire.

e. **Considerations.** Direction, elevation, mask clearance, and adjustment of fire are four essential elements that must be considered when using the MK 19 in defilade position.

(1) ***Direction.*** There are four ways to lay the gun for direction.

(a) *Direct-alignment Method.* The observer selects a position on the gun-target line from which the target can be seen. The gunner loosens the traversing slide and, directed by the observer, moves the MK 19 right or left until it is aligned on the target.

(b) *Aiming-point Method.* For an aiming point, the gunner selects a prominent landmark visible through the sights (Figure 5-22). An aiming point on the gun-target line and at an equal or greater range than the target is desirable; however, the gunner may use an aiming point on the mask.

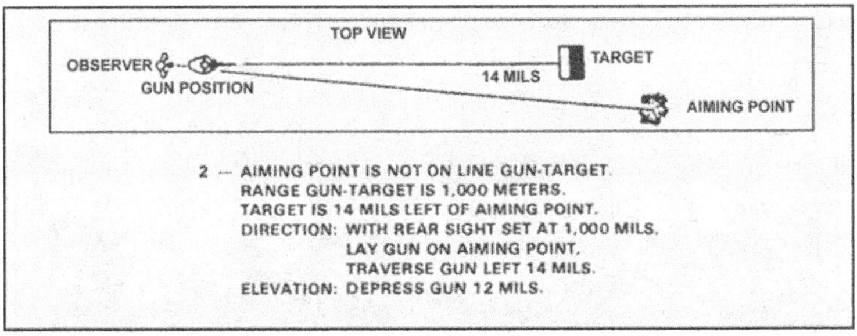

Figure 5-22. Aiming-point method.

- When the aiming point is on the gun-target line, the gunner lays the MK 19 on the aiming point, which aligns it on the target.
- When the aiming point is not on the gun-target line, the gunner measures the deflection with binoculars or compass, and adjusts the lay of the MK 19.

(c) *Aiming-stake Method.* When no natural aiming point is available, the assistant gunner sets out an aiming stake, and the gunner aligns the gun on the target.

(d) *Map-and-compass Method.* The leader locates the MK 19 position and target on a map; draws a line between the two points; orients the map to the terrain; and places the line of sight on the compass along the gun-target line drawn on the map. The leader then announces the magnetic azimuth at the compass index to the gunner as the direction of lay (Figure 5-23). Using this method with terrain-profiling techniques permits the MK 19 to be used in various defilade positions.

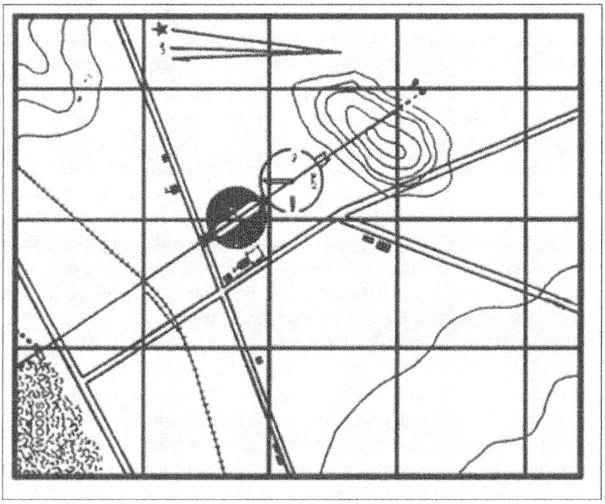

Figure 5-23. Map-and-compass method.

(2) ***Elevation.*** To lay the gun for elevation when engaging targets from defilade, knowledge of the trajectories of 40-mm rounds and of indirect lay is necessary. The MK 19 mounted on the M3 tripod with the T&E mechanism is not equipped to deliver indirect fire in the traditional sense like a mortar. Because the MK 19 has a high-arc trajectory when in the direct-fire mode, it can be fired effectively from defilade if the gunner positions it properly with the help of a well-trained observer. Discussion of laying the MK 19 for elevation includes those techniques and procedures that do not require a lot of data and calculations.

(a) The AE is the vertical angle between the bore line and the line of sight, when the gun and target are at the same elevation (Figure 5-24). The AE is always a positive (plus) and increases as the range increases. The AE for 40-mm ammunition, for each 100 meters of range up to 2,700 meters, is contained in Appendix F, Firing Table. For example, to hit a target at a range of 1,000 meters with M430 HEDP ammunition, the MK 19 must have an AE of +131.9 mils to the line of sight. Is a discussion on leveling the tripod necessary or can the gunner "eye-ball" it?

(b) When the MK 19 and target are not at the same elevation, an additional angle must be taken into consideration. The angle of sight (AS) is the vertical angle formed by the line of sight and a horizontal line from the base of the gun. When the target is at a higher elevation than the gun, the AS is positive (plus) (Figure 5-24). When the target is lower than the gun, the AS is negative (minus).

(c) The angle of quadrant elevation (QE) is formed between the bore line and the horizontal line through the base of the gun (Figure 5-24). The QE is positive (plus) whenever the gun is aimed above the horizontal, and negative (minus) whenever the gun is aimed below the horizontal. It is the algebraic sum of the AE and the AS; that is, if the angle of sight is positive, it is added to the AE; if the AS is negative, it is subtracted from the AE.

QE = AE + AS (target above horizontal line)
QE = AE - AS (target below horizontal line)

Other methods of elevation are as follows:

- **Computed quadrant elevation method.** The leader must determine the correct range to the target. Using the range, the leader finds the corresponding AE from Appendix F, Firing Table. The leader must find the AS using binoculars, by measuring in mils the vertical interval between the target and the estimated horizontal. The leader may assume the distant horizon to be at a zero AS, or at the same elevation as the MK 19 position. QE may be determined by algebraically adding this data as previously described.
- **Measured quadrant elevation method.** The gunner should locate the MK 19 in partial defilade and lay it on the target using direct-lay methods. The leader measures the QE with the M2 compass. The gunner moves the MK 19 into defilade position and places the measured QE on the gun. For each meter difference in elevation between the position in partial defilade and the firing position, the gunner adds 1 mil to the QE when firing at a range of 1,000 meters; 1/2 mil when firing at 2,000 meters, and so on.
- **Aiming-point method.** The gunner selects an aiming point visible from the MK 19 position; preferably a point at a greater range than the target and at a higher elevation than the target, and the leader finds the range to the target. Using binoculars, the leader measures the vertical angle in mils from the aiming point to the base of the target. The leader has the gunner lay the MK 19 on the aiming point, with the sight set to hit the aiming point, and then directs the gunner to manipulate the gun through the number of mils measured from the aiming point to the target. For example, the range to the target is 1,000 meters (Figure 5-25). The angle read with the binoculars from the aiming point down to the base of the target is 12 mils. The sight should be set at 1,000 meters, with the MK 19 laid on the aiming point, and the muzzle then depressed 12 mils.

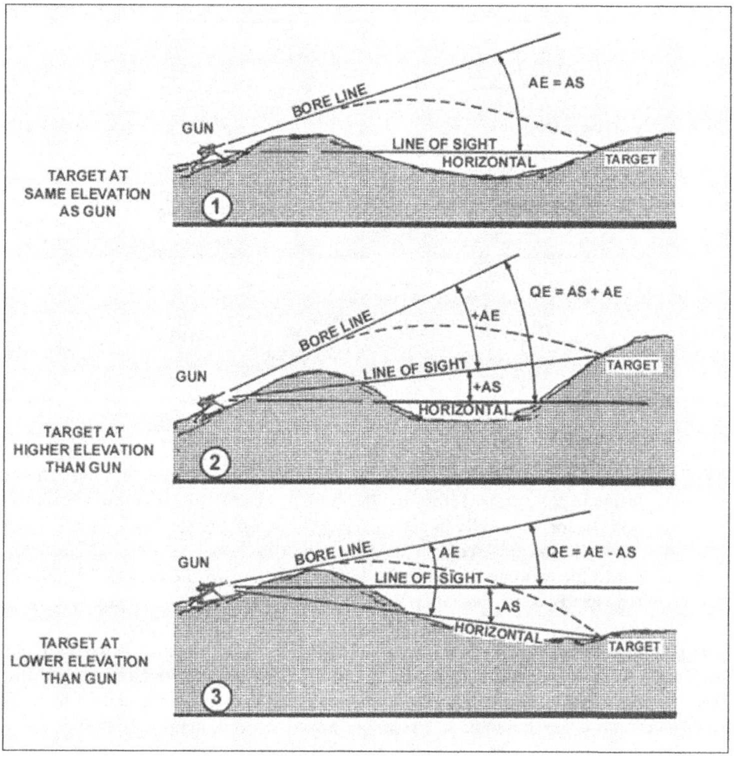

Figure 5-24. Angle of elevation, angle of sight, and quadrant elevation.

(3) ***Mask Clearance.*** After the gun(s) has been laid, the leader must determine whether or not the entire cone of fire will clear the mask, if mask clearance is not obvious. In order to ensure mask clearance, the leader has the gunner sight on the crest of the mask and re-lays the MK 19 for elevation on the target. The gunner ensures that the difference in elevation from the MK 19 to the target and from the MK 19 to the mask is at least +10 mils. The gunner checks this by sighting along the bottom of the receiver and the barrel.

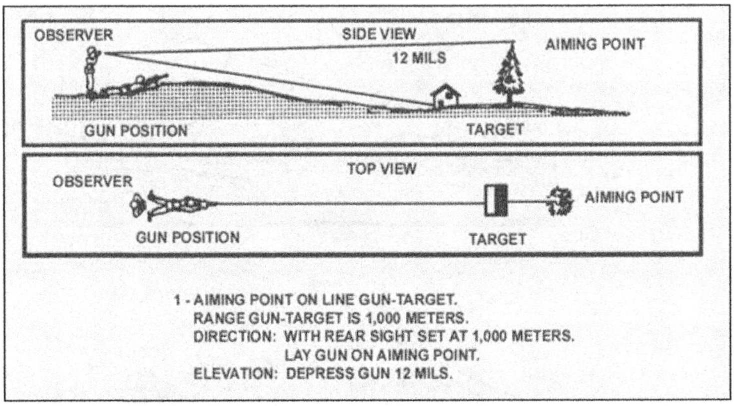

Figure 5-25. Aiming-point method for elevation.

NOTE: The procedures already discussed aid in the initial lay of the gun. The best tactical application of fire from the defilade position is in defense when time is not a factor, especially in a reverse slope defense. Leaders should try to register their MK 19s on selected TRPs, as previously discussed. Use trial and error to adjust for accurate fire from the defilade position. While the MK 19s are being registered, every precaution must be taken to guarantee troop safety.

(4) ***Adjustment of Fire.*** Under field conditions, the MK 19 that is quickly laid on the target seldom results in an initial impact directly on the target. Rapid adjustment of fire is essential. This is assured by thoroughly training the observer to estimate range and lateral distance. The observer should be in a position to see the initial impact. To ensure this, the burst (three to five rounds) should be long. Except for long ranges, or when for other reasons visual observation is hard, the observer does not ordinarily use binoculars for the initial burst. If the initial lay is incorrect, the binocular's limited field of view can cause the observer to miss the impact of the round. Adjustment of fire should be bold and aggressive. The observer should command large corrections and avoid creeping fire toward the target. When a burst is incorrect for deflection and elevation, the observer's next command should correct both.

f. **Construction of a Terrain Profile.** The leader may use a side view or cross section of the ground along a selected line or direction to determine where friendly and enemy forces can see each other. The leader can also use it to plan fires; that is, to locate dead space, to plan a defilade position, and so on. To construct a terrain profile, locate the gun position on the map and determine the direction of fire, or profile line (Figure 5-26).

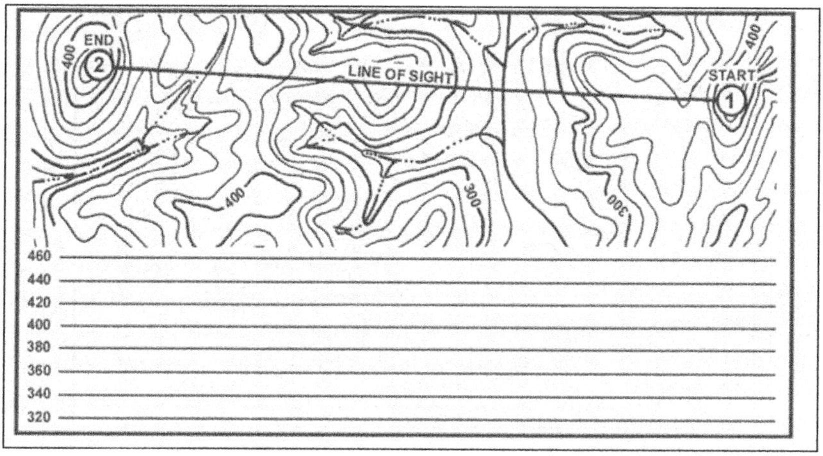

Figure 5-26. Profile line (map contour lines are 20 meters).

(1) Place the edge of a lined piece of paper along the profile line. Tick-mark all points of known elevation on the edge of the paper. These points are the contour lines. Any paper with evenly spaced horizontal lines, such as graph or notebook paper, may be used. The wider the spacing of the lines, the greater the vertical exaggeration in the profile; however this does not affect the information.

(2) Draw perpendicular lines, down across the horizontal lines for each marked point, and identify the high and low points along the profile (Figure 5-27).

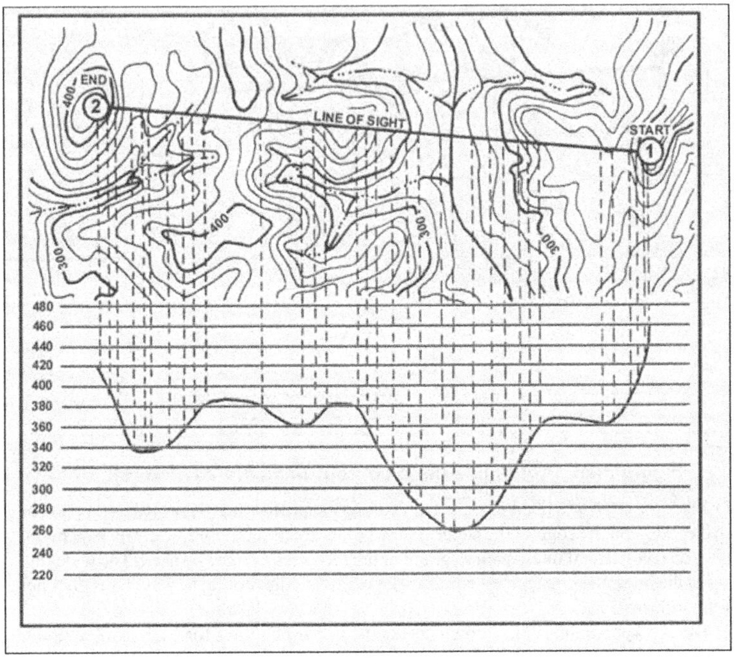

Figure 5-27. Marking of elevation.

(3) Connect all the points with a smooth curve (Figure 5-28).

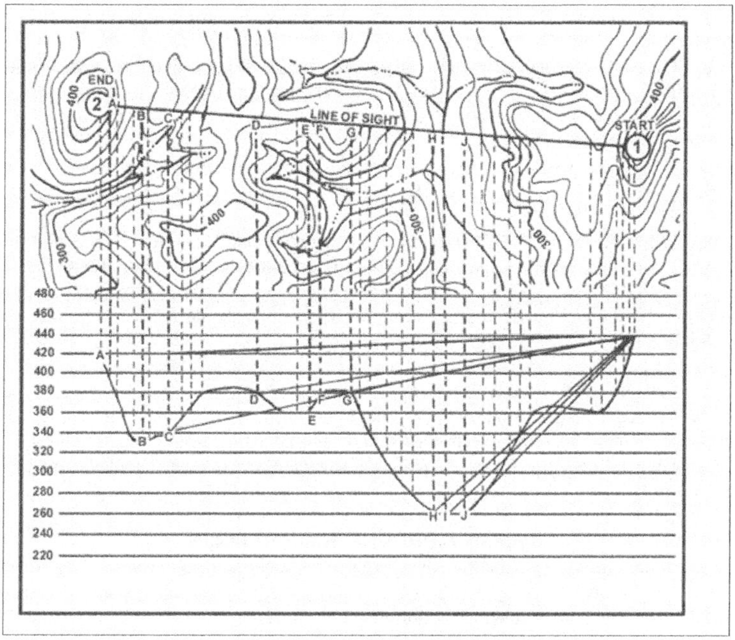

Figure 5-28. Terrain profile.

Section IV. DEGRADED CONDITIONS

The MK 19 is fired the same way under degraded conditions as it is fired under good conditions, except that degraded conditions limit the gunner's ability to observe the impact of rounds in relation to the target and to make adjustments. Refer to Appendix G, Aiming Devices, for additional information.

5-14. NIGHT FIRING

The use of range cards and predetermined fires are important at night or during degraded conditions because the targets or engagement areas are chosen and marked in advance, which enables the gunner to adjust onto targets quickly and easily. The gunner must know how to read his traversing slide bar and T&E mechanism. This task can be difficult at night or during degraded conditions. The gunner can mark targets on the slide bar with luminous tape, use mini-chemical lights, or use flashlights with red lens covers.

a. Firing with the use of illumination has proven to be much the same as firing during normal daylight conditions. Estimation of distances and adjustment of rounds are difficult at night or during degraded conditions. Gunners should be able, using illumination, to effectively engage targets out to 800 meters as well as they can in the daylight. Due to long shadows and dispersion of light past this distance however, visibility is poor even with illumination.

b. The AN/TVS-5 can be mounted on the MK 19 and used effectively to spot and adjust rounds to targets out to ranges of 800 meters. Past this range, the elevation of the scope on the gun is too high to see the impact of the rounds, so it can be handheld by an observer who calls out adjustments to fire.

5-15. FIRING IN A NUCLEAR, BIOLOGICAL, OR CHEMICAL ENVIRONMENT

Firing the MK 19 during NBC conditions is no different than firing in normal conditions. Wearing the protective mask during firing does not change the way the gunner sights and aims the MK 19 to engage targets. The assistant gunner should take care to avoid catching his clothes or gloves while feeding ammunition. The gunner should also avoid contact with the hot barrel.

5-16. FIRING IN FOG OR SMOKE

Neither the degraded conditions of fog or smoke change the way the gunner fires the MK 19. The only limitations are observation of rounds and adjustment onto targets. The use of predetermined fires, trip flares, PEWS, or forward observers can cut down on these limitations.

Section V. PREDETERMINED FIRES

Predetermined fires are planned to cover target areas such as enemy avenues of approach, likely sites for enemy guns, and probable enemy assault routes. Laying the MK 19 on predetermined targets by either of the following methods may be verified by firing the MK 19 and adjusting it on target.

5-17. T&E MECHANISM METHOD

When the MK 19 is laid on target in the primary sector, the direction and elevation is taken from the traversing bar and the T&E mechanism. Both direction and elevation, as well as the range to each target, are recorded on the range card. Each target in the sector of fire must be numbered and laid on in sequence.

5-18. FIELD-EXPEDIENT METHOD OF LAYING THE MK 19

Another method for laying the MK 19 on predetermined targets is to use field expedients (Figure 5-29). Field expedients must be used in the secondary sector, and consist of aiming and base stakes. They can be used in the primary sector to aid the gunner.

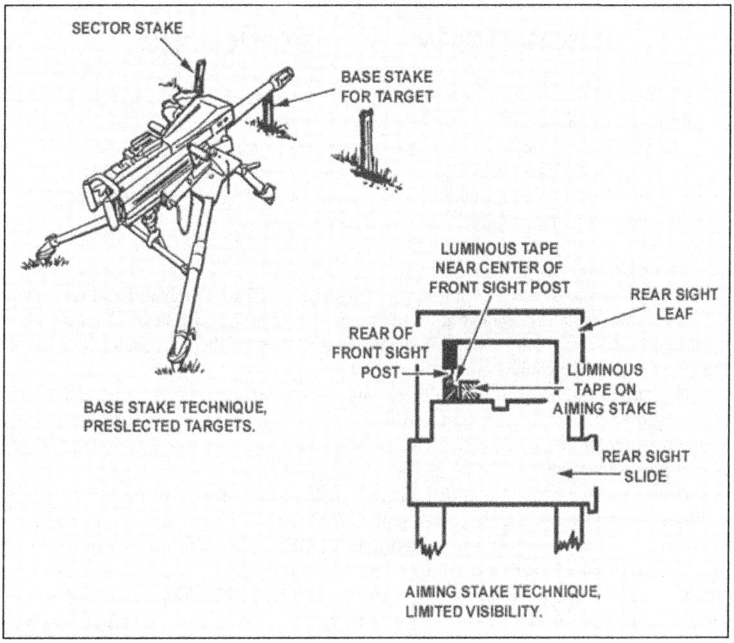

Figure 5-29. Techniques for laying the MK 19 using field-expedient method.

CHAPTER 6

INFORMATION FOR INSTRUCTORS

This chapter provides information to aid instructors in training MK 19 crews to become skilled teams. This chapter is organized to lead the trainer in a progressive manner through the material necessary to aid him in training units. Information is presented beginning with preliminary subjects; these areas include mechanical training, knowledge of the gun's capabilities, and the principles and fundamentals of MK 19 gunnery.

6-1. PRESENTATION

To present instruction in the time allotted, the instructor must be prepared and rehearsed. Assistant instructors and demonstrators must also be trained and rehearsed so they can conduct demonstrations using training aids that clearly illustrate the information. Use public address systems if the group is larger than a platoon.

a. **Organization.** Divide the soldiers into groups of appropriate size, while also considering the phase of instruction and the facilities available. If available, assign an assistant instructor to each crew; otherwise, position the assistant instructors so that they can closely supervise the crew.

b. **Presentation.** Explain and demonstrate each portion of the instruction, following with closely supervised practical work. Emphasize practice throughout training in order to develop skill and teamwork. Secure and critique all MK 19 firing during training.

c. **Examination.** Ensure the soldiers make good progress throughout each period of instruction. Those who do not progress well must receive more instruction under close supervision. Give a formal or informal participation-type examination following each phase of instruction to measure effectiveness of instruction.

6-2. MECHANICAL TRAINING

During mechanical training, the gun is mounted on the M3 mount, or installed on a table. Nomenclature charts (or mats) help the crewmembers learn the names of parts. If possible, assign one assistant instructor for each crew. Present a brief history of the gun. Stress the gun's combat role and assigned mission. Emphasize the purpose, scope, and importance of the instruction to be presented. Be sure to clearly define the term stoppage during the introduction, and emphasize the need for correct care and cleaning procedures.

a. **Organization.** Organize the class into groups of appropriate size. If enough assistant instructors are available, assign one to each crew; otherwise, position the assistant instructors so they can closely supervise the crew.

b. **Presentation.** Table 6-1 can be used as a guide in presenting instruction in stoppages and immediate action. Stress precision in detecting the cause and reducing stoppages. As the soldiers progress, concentrate on speed in applying immediate action and other methods of reducing a stoppage. Train the crew to function as a team.

STOPPAGES	INSTRUCTION PREPARATION
Misfire due to defective primer.	Place a dummy cartridge in the belt.
Short round.	Place a short round in the belt.
Bulged round.	Insert a bulged round in the belt.
Belt loaded improperly.	Pull a cartridge partly out of the belt.
Battered or thick-rimmed cartridge.	Place a battered or thick-rimmed cartridge in the belt.
Failure to remove a round from the chamber.	Place a dummy cartridge, with the rim filed off, in the chamber.
Separated case that is removed from the chamber by the new round when the bolt is pulled.	Drive the front portion of a cartridge securely on a dummy cartridge. Pull the bolt to the rear and place the cartridge properly on the face of the bolt. Ease the bolt forward.

Table 6-1. Preparation for instructing stoppages and immediate action.

6-3. CREW DRILLS

The demonstration crew must be well-trained and equipment must be complete. Use the following equipment for each crew: basic table of organization and equipment (TOE) and distinctive helmet markings that identify each crewmember according to his duties, and normal individual equipment. For more information on crew drills see Appendix H.

a. **Organization.** Present initial instruction and demonstration to one crew. For practical work, divide the class into crews of four men for each MK 19. To aid in control and supervision, mark off equipment with engineer tape or wire, and line up the crewmembers behind their equipment.

b. **Presentation.** Stress the fact that a crew drill is the first step in developing teamwork, and that precision is required of every member for the crew to function smoothly. During initial practice periods, work on precision. Talk each soldier through his duties as a crewmember. Require the soldiers to do each operation when directed with speed and precision.

c. **Advanced Training.** In advanced training, use a speed drill to create interest and stimulate competition. This tactic also furthers teamwork and tests individual crew performance. A well-trained crew should be able, without error, to place the gun into action, or take it out of action in 25 seconds.

6-4. MARKSMANSHIP

With gun and mount, the instructor makes arrangements for a demonstration crew to be available. He ensures that he has a sighting bar, an illustration of a correct sight picture, a sighting target (panel with holder), a stopwatch, control flags, cleaning rods, and patches.

a. **Organization.** Organize the crew and the training area. Organize each crew with four members, if possible. Use a centralized area, preferably with bleachers, for conferences and demonstrations. The area should be fairly level, and large enough for the guns to be placed 10 meters from the targets.

b. **Presentation.** Ensure marksmanship basics are learned and practiced. Make sure the soldiers know that marksmanship basics must be learned and practiced during the preparatory phase of the marksmanship training, in order for them to satisfactorily complete the marksmanship course. With the help of coaches and assistant instructors, ensure that steps are done correctly during the preparatory phase of marksmanship. Give a brief description of the operation, general data, and exterior nomenclature of the MK 19. Ensure that assistant instructors disassemble and assemble the MK 19 while the procedure is being explained.

NOTE: When the crews can skillfully disassemble and assemble the gun, demonstrate this phase of instruction to them blindfolded, using an assistant. Doing this encourages crews to practice during their free time, and develops individual skill and initiative.

Ensure the crew does not assemble or disassemble the MK 19 under stress against time because there is a danger of damaging the gun's parts.

c. **Operation.** Each crew requires a MK 19 mounted on an M3 mount and a belt of five dummy rounds. Emphasize the safety factors involved. Stress how the feed throat must be attached before loading the MK 19. Demonstrate the method of letting the bolt go forward.

d. **Functioning.** Training aids are useful if the class is platoon-sized; otherwise, training aids may be used for study and discussion during breaks. Use one gun for each crew, as in previous mechanical training instruction. Divide functioning into six phases (see Chapter 2, Operation and Function). Explain and demonstrate each phase of functioning. Assistant instructors should duplicate each demonstration on the guns within each crew. With these setups at the guns, each crewmember can see how the parts work. Setups may be used for each phase of functioning.

APPENDIX A
TRAINING STRATEGY

This appendix discusses the two primary components of training strategies: initial and sustainment training.

A-1. COMPONENTS

Training strategies have two primary components: *initial* and *sustainment* training. Both may include individual and collective skills, but *initial training is critical.* With initial training a soldier retains a correctly taught task that he has learned well. With such training, he can easily sustain or quickly regain his performance. Thus, leaders must conduct initial training correctly and well.

a. Trainers use standard, common, and current procedures. Regardless of how well skills are taught or learned, individual and unit skills decay over time. How fast and how much skills decay depends on many factors. Personnel turnover is a primary factor in the decay of collective skills. Other factors include the difficulty and complexity of the tasks.

b. When the unit loses critical crewmembers, it must quickly train new gunners to restore unit proficiency. If too much time passes between initial and sustainment training periods or if the unit alters the established training doctrine, skills erode and the unit might have to completely retrain its MK 19 gunners.

c. A critical step in the Army's overall training strategy is to train leaders and trainers first. Leader courses and unit publications help develop officer and noncommissioned officer (NCO) proficiencies needed to plan and conduct marksmanship training. These resources also help leaders learn how to evaluate the effectiveness of unit marksmanship programs. Proponent schools provide training support materials. These materials include field manuals, training aids, training devices, simulators, and programs of instruction that serve as the doctrinal foundation and guidance for training the force.

A-2. INITIAL TRAINING

The MK 19 training strategy starts with leader training and goes on to unit training. Because "MK 19 gunner" is a Skill Level 2 position, the MK 19 gunner receives most of his MK 19 training within the unit. The training institution provides the unit with a soldier who has received training in the basics of the MK 19. The unit plans minimal MK 19 training for advanced individual training (AIT) soldiers.

A-3. SUSTAINMENT TRAINING

Basic MK 19 gunnery skills and gunner qualification continues in units. Once soldiers learn individual and crew skills, the unit must help them sustain those skills. To do this, leaders follow the annual gunnery-training program provided in Appendix E, Annual Gunnery Training Program. In addition to sustaining skills, the unit incorporates the MK 19 into collective training exercises such as platoon and squad live-fire situational training exercises (STXs).

a. The strategy for sustaining basic MK 19 gunnery skills includes annual or semiannual instructional and qualification range firing. However, the unit must set up a year-round sustainment program. Key elements of the sustainment program include training the trainers, refreshing preliminary skills, and using devices to remedy the training of soldiers who have trouble qualifying on the MK 19.

b. The unit also trains the gunner in other skills such as night firing; mission-oriented protective posture (MOPP) firing; engaging moving targets, using camouflage, cover and concealment; firing and maneuvering; and preparing or selecting a fighting position. The unit also integrates these skills into tactical training. The references at the back of this manual contain additional guidance on the training for these skills.

c. The unit integrates individual and leader MK 19 tasks into collective training such as squad, section, and platoon drills and STXs as well as company LFXs and field training exercises (FTXs). Organizational mission training plans (MTP) and drill books include the tasks, conditions, and standards for the collective tasks in these exercises. These books also explain how to plan and conduct such collective tasks. Based on the type of organization, the unit evaluates collective exercises in leader and trainer after-action reviews (AARs). Objectively evaluating both individual and unit proficiency indicates the readiness and future training requirements.

d. Trainers schedule live fire after the soldiers demonstrate preliminary skills. For initial fire, the trainer conducts an instructional exercise. This improves the soldiers' ability to acquire targets with the MK 19. Once soldiers understand the gun and demonstrate the skills needed to acquire a target, the trainer conducts additional live-fire training and target-detection exercises at various ranges. To develop proficiency, soldiers have to master targets and scenarios of increasing difficulty. After initial individual training, the trainer conducts qualification fire on a range to evaluate each soldier's skills. When conducted in accordance with Chapter 4, Marksmanship, this evaluation also gives leaders some insight into the unit's skill and training effectiveness.

APPENDIX B

OPERATIONS CHECKLIST

This appendix consists of an operations checklist of procedures to be followed before, during, and after use of the MK 19.

WEAPON PART	BEFORE	DURING	AFTER
1. Bore	Ensure it is clear and clean.	______________	Clean and oil lightly.
2. Moving parts	Oil lightly and test for worn or broken parts. The moving parts should function without excessive friction.	Lubricate working parts. Observe the functioning of the gun to anticipate failures.	Inspect, clean, and oil lightly.
3. Ammunition	a. Ensure the correct type of ammunition is used. b. Ensure the ammunition is clean and dent free. c. Ensure all ogives are tight. d. Have an adequate supply of ammunition on hand.	Keep correctly aligned in the feedtray. Check resupply. Protect from sun, moisture, and dirt. Watch for link stoppage.	Clean, store carefully, and replenish supply.
4. Top cover	Inspect the top cover for dents or damage.	Keep closed and locked down.	Lube properly after cleaning.
5. Line of fire	Ensure the line of fire is clear of all obstructions.	Cease fire if any obstruction appears in the line of fire.	____________

Table B-1. Operations checklist.

APPENDIX C

STOPPAGES, MALFUNCTIONS, IMMEDIATE ACTIONS, AND REMEDIAL ACTIONS

A stoppage is any interruption in the cycle of operation caused by faulty action of the weapon or ammunition. A malfunction is a failure of the weapon to function properly. Neither defective ammunition nor improper operation of the gun by a crewmember is considered a malfunction of the MK 19. This appendix lists the actions to correct common stoppages and malfunctions. For more complete information on troubleshooting common malfunctions and stoppages, refer to TM 9-1010-230-10 Troubleshooting Index.

Section I. STOPPAGES, IMMEDIATE ACTIONS, AND REMEDIAL ACTIONS

A stoppage is cleared through immediate or remedial action. Immediate action is the action taken to reduce the stoppage without investigating the cause. The gunner takes remedial action if immediate action does not end the stoppage. Remedial action has three parts: the gunner first unloads and clears the MK 19, identifies the problem, and then applies corrective action to fix the problem. If the MK 19 still does not fire, or cannot be repaired at the user level, contact the unit armorer.

C-1. IMMEDIATE ACTION

Immediate action is different during peacetime and combat.

a. **Immediate Action During Peacetime and Training.** The gunner does the following immediate action to reduce a stoppage during peacetime.

> **CAUTION**
> Both charger handles must be forward and up for firing. If either handle is down, the gun will not fire.

(1) The gunner ensures the safety of other personnel and clears the weapon of ammunition by performing the following steps in sequence:

(a) Clear the area of personnel.

(b) Pull bolt to the rear. If the bolt does not go to the rear, go to BOLT JAMMED.

(c) Catch the live round as it is ejected.

(d) Push the charging handles forward and up.

(e) Put the gun on S (SAFE) and check for any bore obstructions.

(f) If bore is clear, move safety switch from the S (SAFE) to the F (FIRE) position, and attempt to fire.

(g) If nothing happens, put gun on S (SAFE) and wait 10 seconds.

(h) Pull the bolt to the rear.

(i) Catch the live round as it is ejected.

(j) Open the top cover and clear the ammunition.

(k) Check bore for any obstructions.

(2) Charge the MK 19 in the normal manner and check to see the primary pawls have clicked up behind the cartridge in front of the bolt face and that the secondary pawls have clicked up behind the next round before firing. The feed pawls should click up within the 1-inch of charging handle travel.

(3) If the primary and secondary pawls do not click up within the last 1-inch of charging travel, follow these procedures:

- Turn the MK 19 into unit level maintenance.
- Record ammo lot number, type of ammo, number of rounds fired, serial number of the MK 19 and indicate whether ammo is linked with one-piece or two-piece links.

WARNING

1. Do not use combat misfire procedures during peacetime or training. Serious injury can result if precautions are not observed.

2. Do not relink or fire ammunition that has been cycled through the weapon.

b. **Immediate Action During Combat.** To correct a misfire during combat, perform the following immediate actions:

(1) Press charger handle locks and rotate charger handles down.

(2) Pull charger handles to the rear until the bolt sears.

(3) Push charger handles forward and rotate charger handles up and lock.

(4) Relay and fire.

(5) Turn in live rounds that cycled through the weapon as instructed by current directives.

C-2. REMEDIAL ACTIONS TO CLEAR STOPPAGES

If the weapon still does not fire, the gunner clears the weapon, identifies the problem and then takes corrective action to fix or report the problem.

a. **Bad Ammunition.** Bad ammunition can cause a stoppage and is checked first.

(1) ***Primer on the Ejected Round is Indented.*** Clear the weapon and check to see if the primer is indented. If the primer is indented, it is a bad round and dispose of it as authorized. Reload the weapon and continue with the exercise.

(2) ***The Primer is Not Indented.*** Clear the weapon and check to see if the primer is indented. If the primer is not indented, then the firing pin can be bad. Turn the MK19 into unit level maintenance.

(3) ***Round on Bolt Face.*** Clear the weapon. If the round remains on the bolt face, remove the round and place the weapon on S (SAFE) with the bolt to the rear. Check the primer and see if it is indented. See the aforementioned steps for the corrective action.

b. **Ammunition Jammed in Feeder.** A misfire might occur due to ammunition jammed in feeder. Clear the weapon and put gun on S (SAFE).

(1) ***Feed Throat Improperly Attached.*** Check for proper attachment of the feed throat. If the feed throat is attached incorrectly, then re-attach it and reload. Try to recharge and fire the weapon. If charging is not possible, go to BOLT JAMMED.

(2) ***Rounds Crooked or Not Seated Firmly.*** Clear the weapon. If the rounds are crooked or not seated firmly, remove the linked ammunition from feeder. Ensure the link band is even and adjacent to copper band all around ammo. Reload and continue with the exercise.

(3) ***Broken Link.*** A broken link can cause the weapon to stop firing. Clear the weapon. If the link is inside the weapon, remove the link. If it is still on the round, then remove the round and dispose it as authorized. Reload and continue with the exercise.

(4) ***Link Off Rotating Band.*** Clear the weapon. Remove round from belt, dispose of round as authorized. Reload and continue with the exercise.

(5) ***Female Link Not First.*** A round loaded with the female link not first will cause a stoppage. Clear the weapon. Reload the weapon and ensure that the female link is first.

c. **Bad Firing Pin.** A stoppage might occur due to a bad firing pin or firing pin spring. A possible bad firing pin can be determined while checking for a bad round.

(1) ***Primer on Ejected Round Not Indented.*** Clear the weapon. If the primer on the ejected round is not indented, then the firing pin may be bad. Report the defect to the unit armorer.

(2) ***Firing Pin Tip Fails to Spring Forward.*** Clear the weapon. If the firing pin tip fails to spring forward then the firing pin spring can be bad. Report the defect to the unit armorer.

d. **Defective Feeder or Feed Slide Assembly.** A misfire might occur due to a defective feeder or feed slide assembly.

CAUTION
Clear feeder of ammunition. Make sure gun is on S (SAFE).

(1) ***Broken or Worn Feed Pawls.*** Clear the weapon and check to see if the primary and secondary pawls are broken, worn, without spring action, or the pin is missing or dislodged. If they are, turn the MK 19 into unit level maintenance.

(2) ***Badly Worn or Gouged Link Guides.*** Clear the weapon and check to see if the guides are worn or gouged. If they are, turn the MK 19 into unit level maintenance.

(3) ***Binding Feed Slide Assembly.*** Clear the weapon. Remove the feed slide assembly and tray. Clean, inspect and lubricate feed assembly and tray. If the feed slide assembly continues to bind, report the defects to armorer or support maintenance.

e. **Bolt Won't Pick Up the Round.** A misfire might occur because a bolt won't pick up the round.

(1) ***Charging Handles are Down.*** Clear the weapon. Raise the charging handles before firing. If the weapon still does not fire, report defects to the armorer or support maintenance.

(2) ***Dirt-Clogged, Weak, or Damaged Extractors.*** Clear the weapon and remove obstruction. Clean, lubricate, and reload weapon. Continue with the exercise.

(3) ***Feed Slide Out of Adjustment.*** Clear the weapon. The feed slide is out of adjustment if:

- Round fails to feed.
- Round drops.
- Extractors won't pick up round.
- Round stubs on face of chamber.

Report defects to armorer or support maintenance.

f. **Bolt Drops a Round Before Firing.** A stoppage may occur because a bolt drops a round before firing.

(1) ***Weak or Damaged Extractors or Bolt Fingers.*** Clear the weapon and check for weak or damaged extractors or bolt fingers. Report any defects to the armorer or support maintenance.

(2) ***Binding Receiver Rails.*** A misfire might occur because of binding receiver rails. Clear the weapon and pull the bolt to the rear without stopping or pausing. Ease it forward (holding onto one charging handle while you press the trigger) and check for binding. Place weapon on S (SAFE) and remove backplate pin. Lift up slightly on the backplate assembly and pull the bolt and backplate assembly to the rear. Remove the chargers. Check the charger rails and receiver rails for burrs. Report defects to armorer or support maintenance.

WARNING

Be sure bolt is forward before removing backplate pin assembly. If not, serious injury could result.

CAUTION

When installing bolt and backplate, ensure cocking lever is in the forward position. Damage to equipment could result if cocking lever is to the rear.

g. **Bad Cocking Lever.** Clear the weapon and remove the bolt and backplate assembly. Examine the cocking lever on the left side of the bolt for wear or damage. Report defects to armorer or support maintenance.

h. **Unknown Cause.** If the MK 19 stops firing and the procedures above do not identify and correct the problem, then turn it into the armorer or support maintenance.

Section II. MALFUNCTIONS, EMERGENCY ACTIONS AND REMEDIAL ACTIONS

A malfunction is a failure of the weapon to function properly. Neither defective ammunition nor improper operation of the gun by a crewmember is considered a malfunction of the MK 19. The two most common MK 19 malfunctions are sluggish action and runaway gun.

CAUTION
Do not try to break the ammunition belt; injury could result. Lower one charging handle to stop the gun.

WARNING
1. Before performing any non-firing procedure, ensure the weapon is clear of any ammunition.
2. Ensure all ammunition and non-essential personnel are at least 65 meters to the rear of the weapon.
3. If the bolt jams during firing, do not let the bolt slam forward as the top cover is being opened because a round could fire.
4. Be sure to put bolt in forward position before removing the backplate pin assembly. Serious injury can result if the pin assembly is removed with the bolt to the rear.
5. Be prepared to catch dropped/ejected live round from weapon.

C-3. SLUGGISH OR ERRATIC FIRING

Excessive friction from dirt, carbon buildup, lack of lubrication, or burred parts usually causes sluggish or erratic action. Once the gunner realizes that the gun is operating sluggishly or erratically, he should cease-fire and clear the weapon. The two primary corrective actions are cleaning or turning the weapon into the armorer. There are several reasons why a weapon may exhibit sluggish or erratic firing.

a. **Dirty Bore or Chamber.** A dirty bore or firing chamber can cause sluggish or erratic firing. Clear the weapon and clean its bore and chamber. Reload and continue with the exercise.

b. **Recoil Springs or Guide Rods.** Clear the weapon and remove the bolt and backplate assembly from gun. Push against springs to test for weakness. Note bent rods. Report defect to armorer or support maintenance.

c. **Bolt Sear Timing Adjustment.** Clear the weapon and turn it into direct support maintenance.

C-4. RUNAWAY GUN

A runaway weapon continues to fire after the trigger has been released. Worn parts or short recoil of the bolt assembly may cause a runaway gun. Consider the amount of ammunition left and the type of MK 19 mount used when determining the best way to stop the weapon.

a. **Emergency Action.** This paragraph addresses emergency actions that must be taken to stabilize uncontrolled automatic fire. The following procedures should be taken to control a runaway gun:

(1) If ammunition is not low and the MK 19 is used in the free gun mode, keep rounds on target until the all the rounds on the belt have been fired.

(2) If the MK 19 is mounted on either the M3 tripod or on a vehicle with the T&E mechanism attached, hold the grip with one hand. At the same time, press the charger handle lock and lower one charger handle. This action interrupts the cycle of operation, causing the MK 19 to cease firing. The gunner therefore:

(a) Keeps gun pointed downrange and slightly elevated.
(b) Presses charger handle locks.
(c) Lowers the charging handle(s) so the gun will stop firing.
(d) Places the gun on S (SAFE).

WARNING

Never try to break the ammo belt with your hands. Injury could result. Lower one charger handle to stop gun from firing.

b. Once the gun has stopped firing, clear the weapon and report the condition to the armorer or support maintenance.

C-5. BORE OBSTRUCTION

A bore obstruction is indicated by a muffled sound of round firing, excess smoke out of the chamber, and/or excess debris/gases below the gun.

a. **Emergency Action.** The emergency actions for an obstructed bore are the following:

(1) Place weapon on S (SAFE).
(2) Notify range safety officer (during training).
(3) Depress feed pawls, release ammunition belt, and clear feed.
(4) Move the ammunition belt and can to a safe area.
(5) Remove the empty case catch bag.
(6) Charge gun and hold bolt to rear.

WARNING

Do not relink or fire any ammunition that has been cycled through the weapon.

(7) Holding the bolt to the rear, insert a cleaning rod through the receiver rail to the top of the shell casing and as close to the face of the bolt as possible.

(8) Place left hand underneath as close to the round as possible. Raise cleaning rod upward forcing the round off the bolt face into the hand. Remove round to designated area for explosive ordinance disposal (EOD).

b. **Correcting an Obstructed Bore.** To correct an obstructed bore follow these procedures:

(1) Place selector lever on F (FIRE) and ease the bolt forward.

(2) Remove the backplate pin bolt, backplate assembly, vertical cam assembly, and primary drive lever.
(3) Check for any type of obstruction.
(4) Check and remove any case or round from bolt face.
(5) Insert bore obstruction detector into bore to check for a live round.
(6) Remove obstruction per round removal procedures (for exact bore obstruction instructions using the round removal tool, refer to WP 0014 00).

C-6. GUN FIRES TOO SOON

If the weapon fires too soon, or when the trigger is not engaged, the gunner applies emergency action and the weapon is taken to support maintenance. The gunner must not continue to fire the weapon. The following emergency actions must be taken if a weapon fires too soon:

WARNING

Do not attempt to clear the weapon if the weapon fires too soon. Do not attempt to clear or fire the weapon until it is fixed.

(1) Cease fire.
(2) Place weapon on S (SAFE).
(3) Clear area of personnel and ammunition.
(4) Notify the range safety officer.
(5) Check barrel for lodged round using the bore obstruction detector (BOD).
(6) After proper personnel have cleared weapon, evacuate weapon to support maintenance.

C-7. BOLT JAMMED

The following actions are taken if the gunner cannot pull the bolt to the rear.

WARNING

The following procedures must be performed in sequence to open the top cover. The bolt could spring forward suddenly and fire a round causing severe injury. Be prepared to catch ejected round.

a. **Emergency Action and the Bolt Can be Pulled to the Rear.** The following emergency actions must be taken if the weapon initially jams but the bolt can then be pulled to the rear.
(1) Put gun on S (SAFE).
(2) Press charger handle locks and rotate charger handles down.

(3) Pull charger handles to the rear as far as possible without stopping or pausing, until bolt locks. Ensure bolt will stay to the rear before releasing charger handles.

(4) Maintain rearward pressure on charging handles while assistant lifts top cover.

(5) Insert cleaning rod section through slot in side of receiver. Prepare to the catch ejected, live round.

(6) Raise cleaning rod to force live round down. Catch live round as it is ejected.

(7) Remove ammo belt from feeder.

(8) Reposition ammo belt in feeder.

(9) Put gun on F (FIRE).

(10) Ride the bolt forward by grasping one charging handle and depressing the trigger.

(11) Ensure feed slide assembly is to the left.

(12) Ensure secondary drive lever is engaged with the feed slide pin. If not, engage forked end with feed slide pin.

(13) Close top cover gently.

(14) Charge weapon and attempt to fire.

(15) If bolt still jams, repeat first seven steps. Put weapon on S (SAFE), and evacuate to support maintenance.

b. **Gunner's Actions.** The gunner performs the following action once emergency actions are completed:

(1) Clear the jam and make sure there is no bore obstruction.

(2) Charge the MK 19 in the normal manner, and check to see if the primary pawls have clicked up behind the cartridge in front of bolt face and the secondary pawls have clicked up behind the next round before firing. The feed pawls should click up within 1-inch of the charging handle travel.

(3) Check the ammo link to ensure it is even and touches the copper band all around the ammo. If the primary and secondary pawls do not click up within the last 1 inch of the charging handle then:

- Turn MK 19 into unit level maintenance.
- Record ammo lot number, type of ammo, number of rounds fired, serial number of the MK 19 and indicate whether ammo is linked with one-piece or two-piece links.

(4) If the MK 19 does not require feed slide adjustment and there appears to be no other deficiencies that would prevent the weapon from firing, it should be turned into the armorer or support maintenance.

C-8. SHORT RECOIL

A short recoil occurs when the bolt does not fully return to the rear.

> **WARNING**
>
> **When firing HE or TP ammunition, the gunner should observe downrange and attempt to determine if the round left the barrel and also be alert to these three danger signals:**
> **1. A muffled report from the gun.**
> **2. Smoke and debris from the bottom of the receiver.**
> **3. Failure of the projectile to leave the muzzle.**
> **Any of these three symptoms mean a bore obstruction. Do not attempt to clear a bore obstruction.**

a. **Emergency Actions after a Short Recoil.** The gunner does the following actions after a short recoil:

(1) Place weapon on S (SAFE).

(2) Clear area of personnel and ammunition.

(3) Notify range safety officer.

(4) Pull the charger handles to the rear, without stopping or pausing, until the bolt locks. Ensure that the bolt stays to the rear before releasing the charger handles.

(5) Insert cleaning rod section through slot in side of receiver. Prepare to catch the ejected live round.

(6) Raise cleaning rod to force live round down. Catch the live round as it is ejected.

(7) Raise top cover.

(8) Remove ammo belt from feeder.

(9) Check for bore obstruction using bore obstruction detector.

(10)If bore is obstructed, refer to the section above on clearing bore obstructions or refer to WP 0014 00 for round removal procedures.

(11)If there is no obstruction, reposition belt in feeder.

(12)Ensure feed slide assembly is to the left.

(13)Charge weapon and attempt to fire.

b. If a short recoil occurs again, repeat the first eight steps. Put weapon on S (SAFE) and turn it into support maintenance.

C-9. TOP COVER WILL NOT CLOSE

The gunner checks the following so that the top cover can be closed.

a. **Improper Position of the Feed Slide Assembly.** Move feed slide assembly all the way left. The spring should touch the cover.

b. **Bolt is Locked to the Rear.** Ride the bolt forward.

c. **Misaligned Ammunition.** Ensure rounds are straight and firmly seated in the feeder. Ensure links are evenly aligned in the link guide and on rounds. Clean dirt from the feeder.

APPENDIX D
GUNNERY TESTS

There are three gunnery tests: a diagnostic gunnery test, an intermediate gunnery test, and a gunnery skills test. The instructor is responsible for the conduct of all tests and must ensure that sufficient scorecards or answer sheets are available.

D-1. DIAGNOSTIC GUNNERY TEST

This written test, given before and during preparatory training, evaluates trainees' knowledge of the MK 19. It also serves as an evaluation tool for the instruction. Training units prescribe procedures for the test, and announce scores of those who satisfactorily complete the test. If this test is used, it should be given before range firing. To satisfactorily complete the test, the trainee must earn 80 out of 100 points.

		POINTS
1.	Name the five major assemblies of the MK 19.	5
2.	Explain how to clear and disassemble the MK 19.	15
3.	Explain how to assemble the MK 19.	15
4.	Explain how to perform a function check.	5
5.	Explain how to apply immediate action to a runaway gun.	5
6.	Explain how to apply immediate action when a malfunction occurs.	10
7.	Explain one phase of functioning (directed by the examining officer).	10
8.	Explain why the weapon should not be fired with the top cover assembly open.	10
9.	Explain how to inspect ammunition before loading (belt to have at least one each of the following: short round, bent round, loose bullet, broken link or round not pushed fully into its link).	10
10.	Explain how to load to half-load position.	5
11.	Explain and demonstrate sight alignment.	10
	TOTAL	100

D-2. INTERMEDIATE GUNNERY TEST

This written test, given at the end of preparatory marksmanship training, is used to evaluate the instruction as well as the soldier's gunnery skills. This test has 21 questions: 20 multiple-choice and 1 true or false. Each soldier must correctly answer 17 of the 21 in order to move on to the next phase of training. The instructor is responsible for the conduct of the test. The following is a sample of the standard Intermediate Gunnery Test.

INTERMEDIATE GUNNERY TEST

1. What are the MK 19's five major groups (field strip groups)?
   a. Receiver assembly, top cover assembly, sear assembly, bolt assembly, and feed slide assembly.
   b. Top cover assembly, receiver assembly, barrel, sear assembly and feed tray, and bolt and backplate assembly.
   c. Receiver assembly, feed slide assembly and feed tray, top cover assembly, sear assembly, and bolt and backplate assembly.
   d. Sear assembly with chargers, receiver assembly, feed tray assembly, top cover assembly, bolt and backplate assembly.

2. Where on the weapon is the safety located?
   a. On the right side of the receiver assembly.
   b. On the left side of the receiver assembly.
   c. On top of the receiver assembly.
   d. On the sear assembly at the bottom rear of the receiver assembly.

3. In a nonfiring situation to clear the weapon, in which order are the steps below taken?
   a. 2, 4, 3, 1.
   b. 1, 3, 4, 2.
   c. 3, 1, 2, 4.
   d. 1, 2, 4, 3.

   1. Ensure that the safety switch is in the S (SAFE) position, and the bolt is forward. Open the top cover by rotating the top cover latch counterclockwise. Lift the cover.
   2. Return the bolt to the full forward position. Rotate the charger handles upward.
   3. Inspect the feed tray and extractors on the top of the bolt to make sure no round is in the "pickup" position.
   4. Unlock and pull the charging handles to the rear two to three inches. Inspect the face of the bolt and the chamber.

4. In which substance(s) should you never immerse the bolt and backplate assembly?
   a. Cleaning solvent.
   b. Lubricant.
   c. Bore cleaner.
   d. All of the above.

5. Under which conditions should you apply a generous second coat of lubricant when cleaning the weapon?
   a. Hot, wet, salt air, sea spray.
   b. Hot, dry, sandy, dusty.
   c. Cold, icy, snow.
   d. All of the above.

6. Under which conditions should you use a second coat of lubricant?
   a. Hot, wet, salt air, sea spray.
   b. Hot, dry, sandy, dusty.
   c. Cold, icy, snow.
   d. All of the above.

7. During reassembly, what is the first major group you should place back onto the weapon?
   a. Sear assembly.
   b. Bolt and backplate assembly.
   c. Top cover assembly.
   d. Receiver assembly.

8. To ensure the primary and secondary drive levers align properly when you close the top cover of the weapon, in what positions should you place the feed slide assembly and bolt?
   a. Left, to the rear.
   b. Right, to the rear.
   c. Left, forward.
   d. Right, forward.

9. What is the maximum range of the weapon, in meters?
   a. 1,500.
   b. 2,212.
   c. 2,000.
   d. 1,800.

10. What is the maximum effective range of the weapon, in meters, for point targets?

a. 1,500.
b. 2,212.
c. 2,000.
d. 1,800.

11. What is the cyclic rate of fire for the weapon, in rounds per minute?
    a. 225 to 275.
    b. 275 to 325.
    c. 325 to 375.
    d. 375 to 425.

12. What is the main characteristic(s) of the MK 19?
    a. Fully automatic.
    b. Blowback operated.
    c. Air-cooled.
    d. All of the above.

13. Which mount(s) should you avoid attaching the weapon to?
    a. M3 tripod without the T&E mechanism.
    b. M3 tripod with the T&E mechanism.
    c. M4 pedestal.
    d. All of the above.

14. How much armor plating, in inches, can a M430 HEDP round penetrate when fired at zero degrees obliquity?
    a. 1.
    b. 2.
    c. 3.
    d. 4.

15. Within how many meters must the 40-mm HE round impact to cause casualties to exposed enemy personnel?
    a. 5.
    b. 10.
    c. 15.
    d. 20.

16. What steps must you take to stop a runaway weapon?
    a. Release the trigger and step away from the weapon.
    b. Grab and twist the ammunition belt to break it off.

c. Unlock and lower one charging handle while keeping control of the weapon.
d. None of the above.

17. When loading the weapon, place the female link end of the ammunition belt through the feed throat and draw the ammunition into the weapon feed area.
    a. True.
    b. False.

18. What term describes an interruption in the functioning cycle caused either by faulty action of the gun or by faulty ammunition?
    a. Blockage.
    b. Breakage.
    c. Malfunction.
    d. Stoppage.

19. What steps must you take to perform corrective action?
    a. Unload and clear the weapon.
    b. Inspect the weapon and ammunition to find the cause of the problem.
    c. Correct the problem, reload, and try to fire the weapon.
    d. All of the above.

20. When you engage a target, how many rounds should you fire in each burst?
    a. 3 to 5.
    b. 5 to 8.
    c. 7 to 10.
    d. 9 to 12.

21. In which position(s) must you place both charger handles before the weapon will fire?
    a. Forward and down.
    b. Forward and up.
    c. To the rear and down.
    d. None of the above.

D-3. GUNNERY SKILLS TEST

This is a guide for unit commanders to standardize and maintain MK 19 gunnery skills. It is a nonfiring, hands-on test given at the end of MK 19 training. The commander may authorize indoor testing if the facilities are available. Seven stations are used for the skill tests. The test is graded on a GO (pass) or NO-GO (fail) basis.

a. **Instructions for the Gunner's Examination.** The following is a guide for giving gunners the skill test.

(1) To organize for testing, assemble the entire unit in one area and orient them as follows:

> "During the next _____ hours, you will take part in a performance-type examination to test your knowledge of the MK 19 machine gun. Stay with your assigned group during the entire examination. Initially, we will assign each group to a specific station. At each station, an assistant instructor will explain exactly what task or tasks you will perform. After beginning the task, if you do not know a step or steps, ask the assistant instructor for help. He will tell you how to perform the step. He will deduct the necessary points from your grade, and then you may continue to work. You must achieve a score of seventy percent in order to pass the test."

(2) Organize the unit into seven equal groups that rotate from station to station until all individuals have been tested. Gunners must wear their helmet and load-bearing equipment (LBE) to be tested. Assign enough assistant instructors to each station to grade and critique each gunner's performance. For ease in marking, give score sheets to each grader and collect the sheets immediately after the test.

(3) At the end of testing for each group, have each grader assemble the soldiers and give them a thorough five-minute critique.

b. **Testing Stations.** There are seven stations for gunnery skills testing. All stations except one, six, and seven have intermediate training objectives.

c. **Scoring.** For tasks at stations one, six, and seven, score the soldier GO if all steps are passed (P). Score the soldier NO-GO if any steps are failed (F). If the soldier fails any steps, show him what was done wrong and how to do it correctly. For stations two through five, score the soldier GO on a task if he receives no more than one intermediate training objective NO-GO.

STATION ONE

TASK: Identify the characteristics of the MK 19.

CONDITIONS: In any environment, the gunner is given a complete tripod-mounted MK 19.

STANDARDS: The gunner will know characteristics of the MK 19.

PERFORMANCE MEASURES	GO	NO-GO
1. General description (nomenclature).		
2. Weapons sight.		
3. Weight of gun.		
4. Maximum effective range for a point target.		
5. Maximum effective range for an area target.		
6. Cyclic rate of fire.		
7. Muzzle velocity.		
8. Types of ammunition.		
9. Weight of the MK 64, MOD 7, gun mount.		
10. Types of mounts.		

STATION TWO

Station Two has seven intermediate training objectives.

TASK: Perform operator's maintenance on the MK 19.

CONDITIONS: In any environment, the gunner is given a complete MK 19, cleaning equipment, tools, workspace, and TM 9-1010-230-10.

STANDARDS: Disassemble the MK 19 within four minutes. Use the cleaning materials as specified in the operator's TM to clean the MK 19. Inspect, lubricate, and assemble the MK 19 within four minutes, and perform a function check.

Intermediate Training Objective 1: The gunner must disassemble the MK 19.

CONDITIONS: The gunner is given a location in which to work, an assembled MK 19, weapon-cleaning equipment, lubricant, and TM 9-1010-230-10.

STANDARDS: The gunner must disassemble the MK 19 within four minutes.

PERFORMANCE MEASURES	GO	NO-GO
1. Clear the gun.		
2. Remove the bolt and backplate assembly.		
3. Remove the secondary drive lever.		
4. Remove the feed slide assembly.		
5. Remove the top cover assembly.		
6. Remove the feed tray.		
7. Remove the primary drive lever and vertical cam.		
8. Remove the alignment guide assembly.		
9. Remove the ogive plunger.		
10. Remove the round-positioning block.		
11. Remove the charger assemblies (left hand [LH], right hand [RH]).		
12. Remove the sear assembly.		
13. Complete within four minutes.		

Intermediate Training Objective 2: Clean the MK 19.

CONDITIONS: The gunner is given a disassembled MK 19, with the proper cleaning equipment.

STANDARDS: The gunner must clean the MK 19.

PERFORMANCE MEASURES	GO	NO-GO
1. Receiver assembly: a. Wipe or brush away dirt from all parts, especially the interior of the receiver housing, receiver rails, and feeder area. b. Use a rag to apply the cleaning solvent. c. Swab out the bore and chamber using a bore brush and RBC. d. Wipe dry. 2. Sear assembly: a. Wipe or brush away dirt. b. Use a rag to apply the cleaning solvent. c. Wipe dry. 3. Alignment guide assembly: a. Wipe or brush away dirt. b. Soak assembly in cleaning solvent. c. Wipe dry. 4. Ogive plunger assembly: a. Wipe or brush away dirt. b. Use a rag to apply the cleaning solvent. c. Wipe dry. 5. Round-positioning block: a. Wipe or brush away dirt. b. Soak in cleaning solvent. 6. Charger assemblies: a. Wipe or brush away dirt. b. Use a rag to apply cleaning solvent. c. Wipe dry. 7. Vertical cam assembly: a. Wipe or brush away dirt.		

PERFORMANCE MEASURES	GO	NO-GO
b. Soak in cleaning solvent. c. Wipe dry. 8. Primary drive lever: a. Wipe or brush away dirt. b. Soak in cleaning solvent. c. Wipe dry. 9. Secondary drive lever: a. Wipe or brush away dirt. b. Soak in cleaning solvent. c. Wipe dry. 10. Feed slide assembly: a. Wipe or brush away dirt. b. Use a rag to apply the cleaning solvent. c. Wipe dry. 11. Feed tray: a. Wipe or brush away dirt. b. Soak in cleaning solvent. c. Wipe dry. 12. Top cover assembly: a. Wipe or brush away dirt. b. Use a rag to apply the cleaning solvent. c. Wipe dry. 13. Bolt and backplate assembly: a. Wipe or brush away dirt. b. Use a rag to apply the cleaning solvent. c. Wipe dry. 14. Use only authorized cleaning materials.		

Intermediate Training Objective 3: Inspect the MK 19.

CONDITIONS: The gunner is given a disassembled and cleaned MK 19.

STANDARDS: The gunner must inspect the MK 19.

PERFORMANCE MEASURES	GO	NO-GO
1. Receiver assembly:		
a. Inspect the receiver housing for cracks (even hairline), rust, burrs.		
b. Inspect the receiver rails for burrs, deformity, or unusual wear.		
c. Inspect the feeder pawls for spring action, burrs, or broken pawls.		
d. Inspect the barrel for carbon buildup or pitting in the bore or chamber.		
e. Inspect the flash suppressor for dents, cracks, or loose pin.		
f. Inspect the rear sight for rust, and the sight scale for readability.		
2. Sear assembly: Inspect for burrs, rust, or wear on sear face, or binding in the safety switch movement.		
3. Alignment guide assembly:		
a. Inspect for deformed or cracked spring.		
b. Inspect for cracks around the pin.		
c. Inspect for loose pin.		
d. Inspect for burrs, dents, and cracks.		
4. Ogive plunger assembly: Inspect for cracks in the housing.		
5. Round-positioning block:		
a. Inspect for burrs, dents, or cracks.		
b. Inspect for broken, deformed, or missing springs.		
c. Inspect for bent, cracked, or broken parts.		
6. Charger assemblies (LH, RH):		
a. Inspect for burrs on grooved edges.		

PERFORMANCE MEASURES	GO	NO-GO
b. Inspect for chipped, cracked, broken, or missing handle assemblies.		
c. Inspect for spring tension in the charger handle lock.		
d. Inspect for spring tension in the lock plunger.		
e. Inspect for bent charger housing.		
7. Vertical cam assembly:		
a. Inspect for burrs or buckling.		
b. Inspect for scratches, chips, or aluminum buildup on the chromed edge.		
c. Inspect for lock binding.		
8. Primary drive lever: Inspect for burrs around the pivot post.		
9. Secondary drive lever:		
a. Inspect for burrs, cracks, chips, or buckling.		
b. Inspect for missing retaining pin.		
10. Feed slide assembly:		
a. Inspect for burrs or cracks.		
b. Inspect for broken feed pawls, binding, or no spring action.		
c. Inspect for deformed shuttle-spring tip.		
d. Inspect for loose mounting screws.		
11. Feed tray:		
a. Inspect for burrs or cracks.		
b. Inspect for lack of spring action.		
12. Top cover assembly:		
a. Inspect for burrs, cracks, or rust.		
b. Inspect for loose or missing latch.		
c. Inspect for bent pins, and for missing or broken cross pins.		
13. Bolt and backplate assembly:		
a. Inspect overall for burrs, rust, or cracks.		

PERFORMANCE MEASURES	GO	NO-GO
b. Inspect handles for cracks.		
c. Inspect for broken or bent recoil springs.		
d. Inspect for bent or broken guide rods.		
e. Inspect for missing safety wire.		
f. Inspect for broken or worn cocking lever.		
g. Inspect for cracked or chipped bolt sear.		
h. Inspect for cracked or binding cam followers.		
i. Inspect for cracked or chipped bolt fingers.		
j. Inspect for chipped or broken firing pin tip.		
k. Inspect for chipped, cracked, or missing extractors.		
l. Inspect for bent or cracked backplate or missing retaining ring.		

Intermediate Training Objective 4: Lubricate the MK 19.

CONDITIONS: The gunner is given a disassembled, cleaned, and inspected MK 19, and lubricant and brush.

STANDARDS: The gunner must lubricate the weapon.

PERFORMANCE MEASURES	GO	NO-GO
1. Receiver assembly: Apply a light coat of lubricant on the receiver rails, feeder, and feeder pawls.		
2. Sear assembly: Apply a light coat to all parts.		
3. Alignment guide assembly: Apply a light coat to all parts.		
4. Round-positioning block: Apply a light coat to the springs.		
5. Charger assemblies (LH, RH): Apply a light coat to all parts, especially the grooved edges of the rails.		
6. Vertical cam: Apply a light coat to all parts, especially the chromed edge.		
7. Primary drive lever: Apply a light coat to all parts, especially the pivot posts.		
8. Secondary drive lever: Apply a light coat to all parts, especially the pivot post and forked end.		
9. Feed slide assembly: Apply a light coat to all parts, especially the feed pawls.		
10. Feed tray: Apply a light coat to all parts, especially the pawl and guide rails.		
11. Top cover assembly: Apply a light coat to all parts, especially the latch, the pivot post hole, and the pin holes.		
12. Bolt and backplate assembly: Apply a light coat to all parts, especially the bolt face, bolt sear, guide rods, recoil springs, bolt rails, cam followers, and cocking lever.		

Intermediate Training Objective 5: Assemble the MK 19.

CONDITIONS: The gunner is given a disassembled, cleaned, inspected, and lubricated MK 19.

STANDARDS: The gunner must assemble the weapon in the proper sequence within four minutes.

PERFORMANCE MEASURES	GO	NO-GO
1. Assemble the sear assembly.		
2. Assemble the charger assemblies (LH, RH).		
3. Assemble the round-positioning block.		
4. Insert the ogive plunger assembly.		
5. Insert the alignment guide assembly.		
6. Attach the vertical cam.		
7. Attach the primary drive lever.		
8. Attach the feed tray.		
9. Attach the feed slide assembly.		
10. Attach the top cover assembly.		
11. Attach and engage the secondary drive lever.		
12. Insert the bolt and backplate assembly.		
13. Complete within four minutes.		

Intermediate Training Objective 6: Perform a function check.

CONDITIONS: The gunner is given an assembled MK 19.

STANDARDS: The gunner must perform a function check on the weapon.

PERFORMANCE MEASURES	GO	NO-GO
1. Close the top cover.		
2. Charge the weapon.		
3. Lock the charging handles forward.		
4. Place the gun on S (SAFE).		
5. Press the trigger.		
6. If the weapon fires, notify the supervisor.		
7. If the weapon does not fire, continue.		
8. Move the safety to F (FIRE).		
9. Press the trigger. Bolt should spring forward.		
10. Open the top cover.		
11. Check the tip of the firing pin. It should be exposed.		
12. Move the feed slide to the left.		
13. Close the top cover.		
14. Move the safety to S (SAFE).		

Intermediate Training Objective 7: Clean, inspect, and lubricate the MK 64, MOD 7, gun cradle.

CONDITIONS: The gunner is given a gun cradle, TM 9-1010-231-13&P, and lubricant.

STANDARDS: The gunner must clean, inspect, and lubricate the gun cradle.

PERFORMANCE MEASURES	GO	NO-GO
1. Clean the gun cradle. a. Brush or wipe away dirt. b. Use a rag to apply cleaning solvent. c. Wipe dry. 2. Inspect the gun cradle for: a. Bent or cracked retaining pins. b. Damaged threads on depression stop. c. Damage to mounting surface on the gun cradle. d. Loss of mounting hardware (nuts, bolts, pins, or chains). e. Depression stop in place (training use only). f. Frozen pintle lock. g. Missing cotter pins. 3. Lubricate the gun cradle. Apply a light coat to all pins and pivot points.		

STATION THREE

Station Three has four intermediate training objectives.

TASK: Load, unload, correct malfunctions, and clear the MK 19.

CONDITIONS: The gunner is given a location in which to work, a complete M3 tripod-mounted MK 19, and 10 rounds of linked 40-mm ammunition (dummy).

STANDARDS: The gunner must load, unload, and clear the MK 19, and take immediate and remedial action according to TM 9-1010-230-10 and this training program.

Intermediate Training Objective 1: Clear the MK 19.

CONDITIONS: The gunner is given an M3 tripod-mounted weapon in a designated location and 10 rounds of linked 40-mm ammunition (dummy).

STANDARDS: The gunner must clear the MK 19.

PERFORMANCE MEASURES	GO	NO-GO
1. Place the gun on S (SAFE).		
2. Lock the bolt to the rear.		
3. Remove the live round or spent case from the bolt face.		
4. Catch the live round as it falls out.		
5. Open the top cover.		
6. Remove the rounds from the feeder.		
7. Move the safety to F (FIRE).		
8. Ride the bolt forward.		
9. Move the safety to S (SAFE).		
10. Move the feed slide assembly to the left.		
11. Close the top cover.		

Intermediate Training Objective 2: Load the MK 19.

CONDITIONS: The gunner is given an M3 tripod-mounted weapon in a designated location and 10 rounds of linked 40-mm ammunition (dummy).

STANDARDS: The gunner must load the weapon.

PERFORMANCE MEASURES	GO	NO-GO
1. Place the weapon on S (SAFE).		
2. Ride the bolt forward.		
3. Return the charger assemblies to the original upright position.		
4. Open the top cover.		
5. Slide the rounds, female link first, through the feed throat.		
6. Insert the first round into the feeder and across the first pawl.		
7. Move the feed slide assembly to the left.		
8. Close the top cover.		
9. Charge the weapon.		
10. Lock the charger handles forward.		
11. Move the safety to F (FIRE).		
12. Press the trigger (bolt goes forward).		
13. Move the safety to S (SAFE).		
14. Charge the weapon (live round on bolt face).		
15. Lock the charger handles forward.		

Intermediate Training Objective 3: Correct malfunctions.

CONDITIONS: The gunner is given an M3 tripod-mounted weapon in a designated location, 10 rounds of linked 40-mm ammunition (dummy), and a simulated stoppage.

STANDARDS: The gunner must correct the malfunction.

PERFORMANCE MEASURES	GO	NO-GO
1. In a training situation, wait ten seconds after any stoppage. 2. Charge the weapon and try to fire. 3. If the weapon fails to fire, recharge and try to fire. If the weapon fails to fire (or when told that the gun did not fire) after the second attempt, take remedial action: 4. Unload and clear the weapon. 5. Inspect the weapon and ammunition and try to find the cause of the stoppage. 6. If possible, correct the cause of the stoppage, reload, and try to fire the weapon.		

Intermediate Training Objective 4: Stop a runaway gun.

CONDITIONS: The gunner is given an M3 tripod-mounted weapon in a designated location, ten rounds of linked 40-mm ammunition (dummy), and a simulated runaway gun.

STANDARDS: The gunner must stop the runaway gun.

PERFORMANCE MEASURES	GO	NO-GO
Lower one of the charging handles.		

STATION FOUR

Station Four has four intermediate training objectives.

TASK: Mount and dismount the MK 19 onto one of the following: M3 tripod, M4 pedestal on a 1/4-ton vehicle, HMMWV with weapon platform, or commander's cupola on a M113.

CONDITIONS: The gunner is given a complete MK 19, a complete MK 64, MOD 7, gun cradle, a location in which to work, TM 9-1010-231-13&P, and one or more of the following: M3 tripod, 1/4-ton vehicle with an M4 pedestal, HMMWV with weapon platform, or M113.

STANDARDS: The gunner must mount and dismount the MK 19 on the gun cradle, mount the gun on other mounts as directed. The time limit for moving the MK 19 from one mount to another is four minutes.

TRAINING NOTE: When the MK 19 is mounted on a unit's organic carrier or the M3 tripod, it is called a weapon system. The weapon system is mounted using the three-man concept. The gunner is responsible for ensuring that it is mounted correctly, with the aid of the other two-team members. The steps for securing the MK 19 to the gun cradle are the same. The MK 19 is moved from one mount to another with the MK 64, MOD 7, gun cradle attached.

Intermediate Training Objective 1: Mount the MK 19 on the M3 tripod.

CONDITIONS: The gunner is given a location in which to work, an M3 tripod with T&E mechanism, the gun cradle, and TM 9-1010-231-13&P.

STANDARDS: The gunner must mount the gun cradle on the M3 tripod.

PERFORMANCE MEASURES	GO	NO-GO
MOUNTING:		
1. Set up the M3 tripod.		
2. Secure the MK 64, MOD 7, gun cradle onto the M3 tripod.		
3. Lift the gun into the gun cradle.		
4. Slide the front portion of the receiver into the gun cradle.		
5. Secure the rear of the gun to the gun cradle.		
6. Attach the T&E mechanism to the M3 tripod.		
7. Attach the T&E mechanism to the gun cradle.		
8. Attach the feed throat.		

PERFORMANCE MEASURES	GO	NO-GO
DISMOUNTING: 9. Remove the T&E mechanism retaining pin. 10. Insert the gun cradle stow pin (left side). 11. Unlock the M3 tripod pintle lock release cam. 12. Lift the MK 19, with the MK 64, MOD 7, gun cradle attached, from the M3 tripod.		

Intermediate Training Objective 2: Mount and dismount the MK 19 from the M4 pedestal.

CONDITIONS: The gunner is given a location in which to work, a vehicle with an M4 pedestal mount, an M3 tripod-mounted MK 19, the T&E mechanism, pintle adapter, ammunition container bracket, and tools.

STANDARDS: The gunner must remove the MK 19 from the M3 tripod, mount it on the M4 pedestal within four minutes, and dismount it from the M4 pedestal within four minutes.

PERFORMANCE MEASURES	GO	NO-GO
MOUNTING:		
1. Remove the T&E mechanism-retaining pin.		
2. Insert the cradle stow pin (left side).		
3. Unlock the M3 tripod pintle lock lever.		
4. Move the MK 19 to the M4 pedestal.		
5. Loosen the locking lever on the M4 pedestal.		
6. Insert the pintle adapter into the top part of the M4 pedestal.		
7. Remove the quick-release pin from the pintle adapter.		
8. Tighten the locking lever on the M4 pedestal.		
9. Lift up the MK 19 and insert the gun cradle pintle into the top of the pintle adapter.		
10. Insert the quick-release pin into the pintle adapter from either side.		
11. Attach the T&E mechanism to the gun mount.		
12. Center the elevation handwheel.		
13. Remove the stow pin.		
14. Attach the T&E mechanism clamp to the M4 pedestal.		
15. Attach the two support clamps to the M4 pedestal.		
16. Attach the ammunition container bracket (right side).		
17. Attach the empty case catch bag.		
DISMOUNTING:		
18. Remove the ammunition container bracket (right side).		
19. Remove the pin securing the T&E mechanism to the gun cradle.		
20. Insert the gun cradle stow pin.		
21. Remove the quick-release pin from the pintle adapter.		

PERFORMANCE MEASURES	GO	NO-GO
22. Lift the MK 19 from the M4 pedestal.		
23. Loosen the locking lever on the M4 pedestal.		
24. Lift the pintle adapter from the M4 pedestal.		

Intermediate Training Objective 3: Mount and dismount the MK 19 from the HMMWV weapon platform.

CONDITIONS: The gunner is given a mounted MK 19, a location in which to work, a HMMWV with weapon platform, pintle adapter assembly, T&E mechanism, and TM 9-1010-231-13&P.

STANDARDS: The gunner must remove the MK 19 from its mount and install it on the HMMWV weapon platform within four minutes.

PERFORMANCE MEASURES	GO	NO-GO
MOUNTING:		
1. Remove the MK 19 from its mount.		
2. Loosen the locking screws on the HMMWV pedestal.		
3. Insert the pintle adapter into the HMMWV pedestal.		
4. Tighten the locking screws.		
5. Remove the quick-release pin from the pintle adapter.		
6. Lift the MK 19, with MK 64, MOD 7, gun cradle attached to the pintle adapter and insert the gun cradle pintle into the pintle adapter.		
7. Insert the pintle adapter quick-release pin.		
8. Attach the train and elevating assembly to the gun cradle.		
9. Center the elevating handwheel.		
10. Remove the gun cradle stow pin.		
11. Attach the train and elevating mechanism to the HMMWV pedestal.		
12. Attach the ammunition container bracket.		
DISMOUNTING:		
13. Remove the ammunition container bracket.		
14. Remove the train and elevating mechanism from the HMMWV pedestal.		
15. Remove the T&E mechanism from the gun cradle.		
16. Insert the gun cradle stow pin.		

PERFORMANCE MEASURES	GO	NO-GO
17. Remove the quick-release pin from the pintle adapter assembly.		
18. Lift the MK 19 from the HMMWV pedestal.		
19. Remove the pintle adapter from the HMMWV pedestal.		

Intermediate Training Objective 4: Mount and dismount the MK 19 from the commander's cupola of an M113.

CONDITIONS: The gunner is given a mounted MK 19, a location in which to work, an M113 with the commander's cupola, pintle adapter, ammunition bracket, and TM 9-1010-231-13&P.

STANDARDS: The gunner must remove the MK 19 from its mount and install it on the commander's cupola of the M113 within four minutes.

PERFORMANCE MEASURES	GO	NO-GO
MOUNTING:		
1. Insert the pintle adapter into the mount receptacle.		
2. Remove the pintle adapter quick-release pin.		
3. Lift the MK 19 to the pintle adapter and insert the cradle pintle into the pintle adapter.		
4. Insert the pintle adapter quick-release pin.		
5. Attach the ammunition bracket.		
DISMOUNTING:		
6. Insert the gun cradle stow pin.		
7. Remove the ammunition bracket.		
8. Remove the pintle adapter quick-release pin.		
9. Lift the MK 19 from the commander's cupola.		

STATION FIVE

TASK: Identify characteristics of a defensive position for the MK 19.

CONDITIONS: The gunner is given an M3 tripod, a complete MK 19, a location in which to work, a direction of fire, and a partially dug fighting position.

STANDARDS: The gunner must identify the actions needed to complete the fighting position.

PERFORMANCE MEASURES	GO	NO-GO
1. Check assigned direction of fire.		
2. Check gun traverse.		
3. Check overhead cover.		
4. Check fighting position camouflage.		
5. Check for protection against small-arms fire.		
6. Check for protection against indirect fire fragments.		
7. Check for protection from aerial observation.		
8. Check for mutual support with other fighting positions.		
9. Check for a good field of fire.		

STATION SIX

TASK: Prepare a range card for the MK 19.

CONDITIONS: During daylight, in a defensive position, the gunner is given a tripod-mounted MK 19, a principal direction of fire and secondary sector of fire (both with recognizable targets), paper, pencil, and compass.

STANDARDS: The gunner must prepare range cards that include a data and sketch section for each sector of fire, with appropriate sketches and military symbols.

PERFORMANCE MEASURES	GO	NO-GO
1. Orient the range card.		
2. Sketch the terrain that is to the front of the position (prominent and man-made, including terrain features).		
3. Fill in or construct the marginal data section.		
a. Gun number.		
b. Unit designation.		
c. Date.		
d. Magnetic north arrow.		
4. Sketch machine gun symbol.		
5. Specify gun location in relation to prominent terrain feature or eight-digit grid coordinate.		
a. Determine azimuth in mils from the terrain feature to the gun.		
b. Determine distance between the terrain feature and the gun.		
c. Sketch and identify the terrain feature on the card.		
d. Connect the feature and gun with a barbed line.		
e. Write the distance in meters above the line.		
f. Write the azimuth in mils below the line from the feature to the gun.		
6. Sketch in the TRPs, if any.		
7. Label TRPs, if any.		

STATION SEVEN

TASK: Mount, place into operation, and dismount the AN/TVS-5 NVD from the MK 19.

CONDITIONS: The gunner is given a mounted MK 19 and an AN/TVS-5.

STANDARDS: The gunner must mount the AN/TVS-5 on the MK 19, place the sight into operation, and dismount the AN/TVS-5 from the MK 19.

PERFORMANCE MEASURES	GO	NO-GO
1. Mount the M2 mounting bracket onto the MK 19.		
2. Mount the AN/TVS-5 onto the M2 bracket.		
3. Install the batteries in the device.		
4. Press your eye against the eye guard.		
5. Turn the ON-OFF/TUBE BRIGHTNESS control to ON.		
6. Adjust the ON-OFF/TUBE BRIGHTNESS control, as needed.		
7. Turn the ON-OFF/RETICLE BRIGHTNESS control to ON.		
8. Turn the diopter focus ring until the clearest image of the reticle pattern is obtained.		
9. Turn the objective focus ring until the target in the field of view is sharply defined.		
10. Turn the ON-OFF/RETICLE BRIGHTNESS control to OFF.		
11. Turn the ON-OFF/TUBE BRIGHTNESS control to OFF.		
12. Remove the batteries from the AN/TVS-5.		
13. Remove the AN/TVS-5 from the M2 bracket.		
14. Remove the M2 bracket from the MK 19.		

APPENDIX E

ANNUAL GUNNERY TRAINING PROGRAM

This appendix contains four suggested annual training programs for the MK 19, MOD 3; one for each of the fighting and supporting active and reserve forces. Because of the high cost of M918TP and M430 HEDP rounds, live fire is kept to a minimum. Gunnery training programs have been prepared to adequately train gunners based on the mission profile of the unit. Refer to Chapter 4, Marksmanship, and Appendix I, Gunnery Exercises, for the firing tables that support these programs. A definition of each force, along with a summary of the annual gunnery training program designed for it, is included in this appendix.

E-1. LIVE-FIRE EXERCISES

LFXs allow the gunner to qualify on the MK 19 and to participate in collective gunnery exercises. Gunners practice and qualify using Tables I through IV in Chapter 4. The gun crews practice individual and collective skills using Tables I-1 through I-7 in Appendix I.

a. **Practice and Qualification Table Matrix** (Table E-1)**.** The tables used for gunner practice and qualification are in Chapter 4. There are four scorecards available and they are used based on the type of target (hull or pop-up silhouettes) and whether the practice-qualification is during the day or during limited visibility. Each scorecard has two tables, one for practice and one for qualification. The tables have versions for hull or pop-up silhouette engagements and for the type of NVD used. Sample scorecards are shown in Figures 4-27 through 4-30 (and reproducible forms are provided in the back of this manual). Gunners will only fire one day practice-qualification and one night practice-qualification. Units should select the practice and qualification based on the light conditions, type of targets available, and type of NVDs used. The following table shows which scorecards should be used.

Conditions	Target	Night Vision Devices	Scorecard
Day	Hull	NA	Scorecard I
Day	Pop-Up	NA	Scorecard III
Limited/Night	Hull	**AN/PEQ-2A** mounted on the TWS mounting bracket. **AN/PAS-13** mounted on the TWS mounting bracket. **AN/TVS-5** upgraded with the 3d generation tube mounted on the TWS mounting bracket.	Scorecard II
Limited/Night	Pop-Up or E-Type	All night vision devices.	Scorecard IV
Limited/Night	Any type	No night vision device.	Scorecard IV

Table E-1. Practice and qualification tables.

b. **Gunnery Exercises.** Gunnery exercises allow individual gun crews to practice firing the MK 19 while performing other tasks such as mounting the gun and preparing range cards. They also allow multiple gun crews to coordinate their fires while engaging targets. Table E-1 outlines each of the seven exercises. Based on their unit's training needs and the resources available, commanders must decide on the number and type of gunnery exercises to include in their annual training program.

Table	Exercise	Number of Guns	Number of Rounds
I-1	Instructional Fire Exercise	1	26 rounds
I-2	Dismounted Range Card Exercise	1	42 rounds
I-3	Mounted Combat Exercise	1	26 rounds
I-4	Night Fire Exercise	1	22 rounds
I-5	Linear and Deep Targets	2 to 4	80 rounds per gun
I-6	Linear Targets with Depth	2 to 4	40 rounds per gun
I-7	Area Targets	4 to 6	80 rounds per gun

Table E-2. Gunnery exercises.

E-2. FIGHTING FORCES ACTIVE COMPONENT

These forces consist of infantry, military police, scout, and engineer units of the active Army. Gunners assigned to these units are required to qualify semiannually. They are also required to participate in squad/platoon-level live fires quarterly and combined arms live fire exercises (CALFEX) annually. Assistant gunners are also required to qualify or fire familiarization exercises annually.

a. *Every quarter* the unit conducts preliminary marksmanship instruction and crew drills. The Gunnery Skills Test is also conducted.

b. During the *first quarter*, semiannual gunner practice qualification and qualification exercises are performed (see Table E-1 for the correct tables to use). Assistant gunner familiarization (with a zeroed gun) using Tables I and III are also conducted (see table E-1). Unit LFXs are performed (see Tables I-5 through I-7).

c. During the *second quarter,* unit LFXs are performed (see Tables I-1 through I-4, and I-5 through I-7).

d. During the *third quarter,* semiannual gunner practice qualification and qualification exercises are performed (see Table E-1 for the correct tables to use). Unit LFXs are also performed (see Tables I-5 through I-7).

e. During the *fourth quarter,* conduct LFXs or CALFEX using exercises from Tables I-5 through I-7.

E-3. SUPPORTING FORCES ACTIVE COMPONENT

These forces consist of all other units in the active component not identified as fighting force units. These guns are used primarily in defense of battalion, brigade, and division rear operations. Gunners from these forces participate in instruction fire and qualification annually.

a. *Every quarter* the unit conducts preliminary marksmanship instruction and crew drills. The Gunnery Skills Test is also conducted.

b. During the *first quarter,* there is no additional training.

c. During the *second quarter,* perform LFXs (see Tables I-5 through I-7).

d. There is no additional training during the *third quarter.*

e. During the *fourth quarter,* annual day gunner practice qualification and qualification exercises are performed. Night practice qualification and qualification is conducted once every two years (see Table E-1 for the correct tables to use). Once every two years, assistant gunner familiarization (with a zeroed gun) using Tables I and III is also conducted (see table E-1).

E-4. FIGHTING FORCES RESERVE COMPONENTS

These forces consist of military police, engineer, and infantry units of the reserve components. Gunners for these forces must maintain a level of proficiency higher than that of the active support forces. At the same time, limited time and range resources must be considered. Gunners of these forces conduct sustainment on the trainer at least quarterly. Live fires consist of annual qualification and a unit level LFX. These forces are allocated 192 or 118 rounds each year for live fire and qualification.

a. *Every quarter* the unit conducts preliminary marksmanship instruction and crew drills. The Gunnery Skills Test is also conducted.

b. There is no additional training during the *first quarter.*

c. There is no additional training during the *second quarter.*

d. During the *third quarter,* annual gunner practice qualification and qualification exercises are performed. See Table E-1 for the correct tables to use. Once every two years, the assistant gunner fires familiarization (with a zeroed gun) using Tables I and III (see table E-1). Unit LFXs are performed (see Tables I-5 through I-7).

e. There is no additional training during the *fourth quarter only.*

E-5. SUPPORTING FORCES RESERVE COMPONENT

These forces consist of all reserve units not specifically outlined under the paragraph on fighting forces. They are allocated 78 rounds for each year to qualify.

a. *Every quarter* the unit conducts preliminary marksmanship instruction and crew drills. The Gunnery Skills Test is also conducted.

b. There is no additional training during the *first quarter.*

c. There is no additional training during the *second quarter.*

d. During the *third quarter,* gunner practice qualification and qualification exercises are performed once every two years. See Table E-1 for the correct tables to use. Once every two years, the assistant gunner fires familiarization (with a zeroed gun) using Tables I and III (see table E-1). Unit LFXs are performed (see Table I-7).

e. There is no additional training during the *fourth quarter.*

APPENDIX F

FIRING TABLE

The table in this appendix provides information about the weapon elevation, time of flight, and other factors of MK 19 projectiles.

NOTE: 40-mm TP, XM918 is reduced time-of-flight experimental ammunition.

	ELEVATION (MILS)				FLIGHT TIME (SECONDS)				REMAINING VELOCITY			
									M430		M385	
RANGE (M)	M430	M385	M918	XM918*	M430	M385	M918	XM918*	FPS	MPS	FPS	MPS
0	0.0	0.0	0.0	0.0	0.00	0.00	0.00	0.00	790	240.8	790	240.8
100	9.0	9.0	8.9	4.6	0.43	0.43	0.43	0.31	747	227.7	744	226.8
200	18.6	18.7	18.5	9.5	0.88	0.88	0.88	0.63	706	215.2	701	213.7
300	29.1	29.3	28.8	14.9	1.36	1.37	1.35	0.97	667	203.3	660	201.2
400	40.4	40.8	39.9	20.6	1.87	1.88	1.85	1.33	630	192.0	622	189.6
500	52.6	53.4	51.8	26.7	2.40	2.43	2.38	1.70	596	181.7	586	178.6
600	65.9	66.9	64.7	33.3	2.97	3.07	2.93	2.10	564	171.9	552	168.2
700	80.3	81.8	78.5	40.4	3.58	3.63	3.52	2.52	533	162.5	520	158.5
800	96.0	98.2	93.5	48.1	4.22	4.29	4.14	2.97	505	153.9	491	149.7
900	113.1	116.1	109.8	56.3	4.90	4.99	4.80	3.43	478	145.7	463	141.1
1,000	131.9	135.8	127.5	65.2	5.63	5.74	5.50	3.93	453	138.1	438	133.5
1,100	152.5	157.6	146.8	74.9	6.41	6.55	6.24	4.45	430	131.1	414	126.2
1,200	175.2	181.9	167.8	85.3	7.24	7.42	7.03	5.00	409	124.7	393	119.8
1,300	200.4	209.0	191.0	96.6	8.14	8.37	7.88	5.58	389	118.6	373	113.7
1,400	228.6	239.6	216.6	108.9	9.11	9.40	8.79	6.19	372	113.4	356	108.5
1,500	260.3	274.6	245.0	122.2	10.1	10.53	9.77	6.84	356	108.5	341	103.9
1,600	296.7	315.3	277.1	136.7	7	10.53	10.84	7.53	343	104.5	329	100.3
1,700	339.2	364.2	313.7	152.6	11.3	11.80	12.01	8.27	333	101.5	320	97.5
1,800	390.8	426.4	356.5	169.9	4	13.25	13.34	9.05	325	99.1	314	95.7
1,900	458.3	518.4	408.5	189.0	12.6	15.01	14.88	9.88	322	98.1	314	95.7
2,000	570.1	X	477.2	210.0	5	17.44	16.82	10.77	324	98.8	X	X
2,100	x	X	757.3	233.4	14.1	X	23.66	11.73	X	X	X	X
2,200	X	X	X	259.6	7	X	X	12.77	X	X	X	X
2,300	X	X	X	289.3	16.0	X	X	13.91	X	X	X	X
2,400	X	X	X	323.4	6	X	X	15.18	X	X	X	X
2,500	X	X	X	363.8	18.9	X	X	16.62	X	X	X	X
2,600	X	X	X	413.8	7	X	X	18.32	X	X	X	X
2,700	X	X	X	482.4	X	X	X	20.54	X	X	X	X
2,800	X	X	X	X	X	X	X	X	X	X	X	X
					X							
*REDUCED TIME-OF-FLIGHT EXPERIMENTAL AMMUNITION.												

Table F-1. Comparison of direct-fire aiming data.

	REMAINING VELOCITY				ANGLE OF FALL (DEGREES)				MAXIMUM ORDINATE (FEET)			
	M918		XM918*									
RANGE (M)	FPS	MPS	FPS	MPS	M430	M385	M918	XM918*	M430	M385	M918	XM918*
0	790	240.8	1100	335.3	0.0	0.0	0.0	0.0	0.0	0.0	0.0	0.0
100	750	228.6	1044	318.2	0.5	0.5	0.5	0.3	0.7	0.7	0.7	0.4
200	712	217.0	991	302.1	1.1	1.2	1.1	0.6	3.1	3.1	3.1	1.6
300	675	205.7	940	286.5	1.8	1.9	1.8	0.9	7.4	7.5	7.3	3.8
400	641	195.4	892	271.9	2.6	2.7	2.6	1.3	14.0	14.2	13.8	7.1
500	608	185.3	847	258.2	3.6	3.7	3.5	1.8	23.3	23.7	22.8	11.7
600	578	176.2	804	245.1	4.7	4.8	4.5	2.3	35.7	36.5	34.7	17.9
700	549	167.3	763	232.6	5.9	6.1	5.6	2.9	51.7	53.1	50.1	25.7
800	521	158.8	724	220.7	7.3	7.6	6.9	3.6	72.0	74.4	69.4	35.6
900	496	151.2	688	209.7	8.9	9.3	8.4	4.3	97.4	101.0	93.4	47.7
1,000	471	143.6	653	199.0	10.7	11.3	10.1	5.2	128.7	134.1	122.7	62.5
1,100	449	136.9	620	189.9	12.8	13.6	12.0	6.1	167.0	174.8	158.2	80.3
1,200	428	130.5	590	179.8	15.2	16.2	14.1	7.2	213.6	224.9	201.2	101.6
1,300	409	124.7	561	171.0	17.9	19.1	16.5	8.4	270.3	286.3	252.9	126.8
1,400	391	119.2	534	162.8	21.0	22.5	19.2	9.8	339.3	361.8	315.0	156.6
1,500	376	114.6	508	154.8	24.5	26.4	22.3	11.4	423.6	455.3	390.0	191.6
1,600	362	110.3	484	147.5	28.4	30.9	25.8	13.1	527.6	527.9	480.8	232.7
1,700	350	106.7	462	140.8	32.9	36.1	29.7	15.0	658.4	724.6	592.2	280.8
1,800	341	103.9	442	134.7	38.2	42.2	34.1	17.2	828.4	931.6	731.6	337.0
1,900	334	101.8	423	128.9	44.4	50.0	39.2	19.6	1,065.6	1,259.3	912.3	402.8
2,000	332	101.2	406	123.7	53.2	X	45.3	22.4	1,486.5	X	1,166.1	479.9
2,100	350	106.7	392	119.5	X	X	63.6	25.3	X	X	2,298.7	570.7
2,200	X	X	379	115.5	X	X	X	28.6	X	X	X	678.0
2,300	X	X	368	112.2	X	X	X	32.3	X	X	X	806.2
2,400	X	X	359	109.4	X	X	X	36.3	X	X	X	961.6
2,500	X	X	354	107.9	X	X	X	40.8	X	X	X	1,154.6
2,600	X	X	351	107.0	X	X	X	45.8	X	X	X	1,405.6
2,700	X	X	352	107.3	X	X	X	52.0	X	X	X	1,767.3
2,800	X	X	X	X	X	X	X	X	X	X	X	X

*REDUCED TIME-OF-FLIGHT EXPERIMENTAL AMMUNITION.

Table F-1. Comparison of direct-fire aiming data (continued).

APPENDIX G

AIMING DEVICES

This appendix provides information on the employment of the AN/PEQ-2A target pointer illuminator aiming light (TPIAL), AN/TVS-5, and the AN/PAS-13B (V) 3 heavy weapon thermal sight (HWTS). These devices have the advantage of greater magnification and resolution for the employment of the MK 19 during periods of degraded visibility. These devices, when properly boresighted and employed under proper conditions, allow infantrymen to engage targets at the weapon's maximum effective ranges during periods of degraded visibility.

G-1. AN/PEQ-2A DESCRIPTION AND OPERATION

The TPIAL projects an infrared laser beam that cannot be seen with the eye but can be seen with NVD. It is also capable of projecting a much wider infrared illuminating beam from an integral illuminator. The TPIAL works with night vision goggles and mounts on various weapons with mounting brackets and adapters (Figure G-1 and Table G-1). Leaders can also use the AN/PEQ-2A in the hand-held mode to illuminate and designate targets.

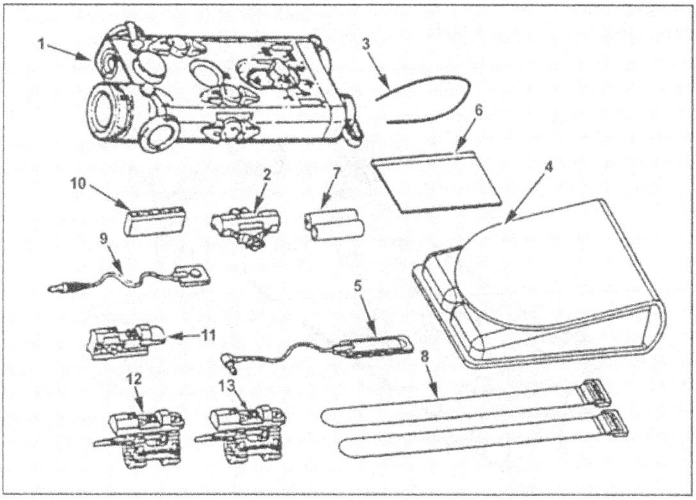

Figure G-1. AN/PEQ-2A major components.

ITEM	DESCRIPTION
1	TPIAL Assembly with Safety Block
2	Rail Grabber Mounting Bracket
3	Neck Cord
4	Carrying Bag
5	Cable Switch, 20 inch Remote, Button
6	Operator's Manual
7	Batteries, 2 AA
8	Straps, Retention
9	Cable Switch, 12 inch, Membrane
10	Bracket Adapter
11	Training Extender (Army only)
12	M4/M16A2 Bracket Assembly (Army only)
13	M16A2 Bracket Assembly (Marine only)

Table G-1. Names of the AN/PEQ-2A major components.

a. **Controls and Indicators.** The AN/PEQ-2A has controls and indicators that allow the user to operate the device and to select its different modes.

(1) ***Battery Installation.*** To install batteries in this device, first unscrew the battery caps and install two AA batteries. Orient the batteries as indicated by the markings on the AN/PEQ-2A body (Figure G-2).

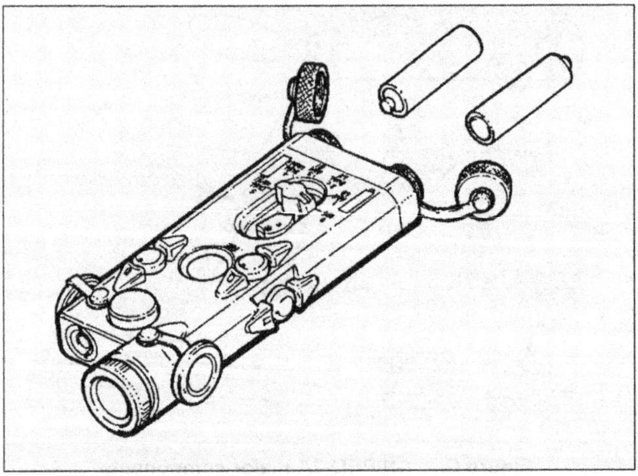

Figure G-2. AN/PEQ-2A battery installation.

(2) ***Safety Block Installation.*** The safety block installed in the training mode (blue side up) prevents the operator from accessing the non-eye safe modes (AIM HI, DUAL

LO/HI, DUAL HI/HI) (Figure G-3). A .050 hex-head Allen Wrench is needed to unscrew the block from the body and re-install it in the tactical mode (black side up).

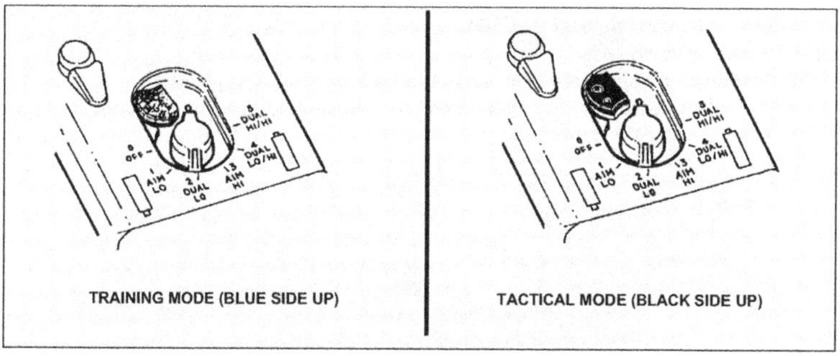

Figure G-3. Safety block installation.

(3) ***Mode Selector.*** The mode selector is used to set the mode in which the AN/PEQ-2A will operate when the cable switch button or push button is depressed. The mode selector has six positions (Table G-2).

KNOB POSITION		OPERATION
0	OFF	The AN/PEQ-2A will not operate.
1	AIM LO	The aiming beam operates at low power.
2	DUAL LO	The aiming beam and the illuminating beam operate at low power.
3	AIM HI	The aiming beam operates at high power.
4	DUAL LO/HI	The aiming beam operates at low power and the illuminating beam operates at full power.
5	DUAL HI/HI	The aiming beam operates at high power and the illuminatlng beam operates at full power.

Table G-2. Mode selector position.

(4) ***Button Switch.*** The button switch is used when the AN/PEQ-2A is hand held. Pressing the button switch operates the AN/PEQ-2A in the operational mode set by the

mode selector switch (Figure G-4). When the button is released, the AN/PEQ-2A turns off.

(a) A green LED is incorporated into the body of the AN/PEQ-2A to indicate that the unit is ON. Whenever the AN/PEQ-2A is activated, the green LED will light and stay lit until the unit is turned OFF.

(b) If continuous operation of the AN/PEQ-2A is desired, pressing the button switch twice in rapid succession will latch the unit ON. The unit will remain ON until the button switch is pressed the third time.

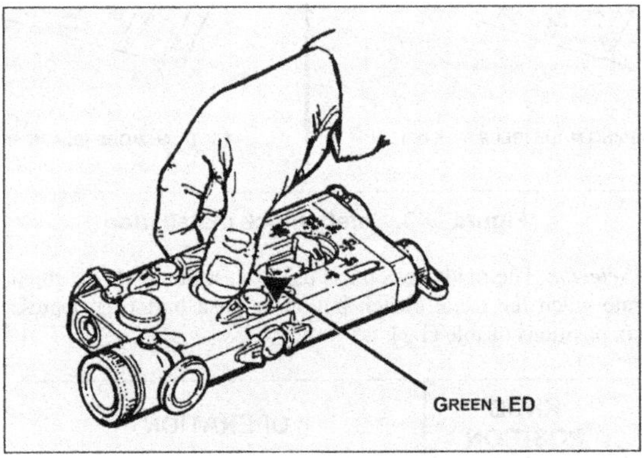

Figure G-4. Operation of the button switch.

(5) ***Cable Switch.*** The cable switch is used when the AN/PEQ-2A is mounted on a weapon (Figure G-5). The cable switch plugs into the back of the AN/PEQ-2A assembly. Pressing the button or pad at the end of the cable switch causes the unit to turn on in the operational mode selected by the mode select switch. When released, the AN/PEQ-2A turns off.

(a) If continuous operation of the AN/PEQ-2A is desired, pressing the cable switch twice in rapid succession will latch the unit ON. The unit will remain on until the push button is pressed a third time.

(b) When the cable switch plug is installed in the AN/PEQ-2A, it automatically locks into place. To remove the switch, pull back on the plug sleeve and pull the plug out.

CAUTION
Do not try to remove the plug by pulling on the cable.

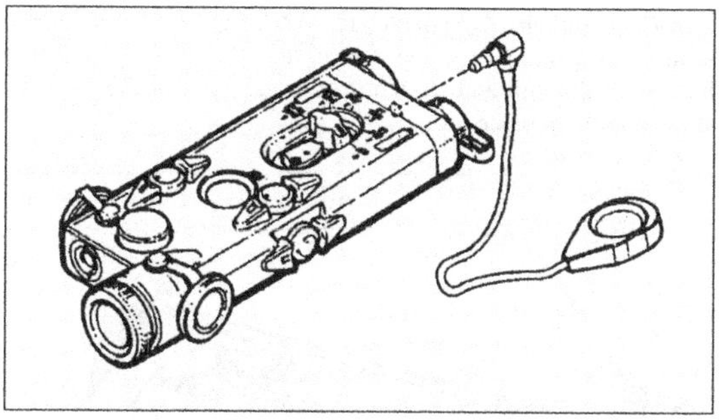

Figure G-5. Installation of the cable switch.

(6) ***Focus Knob.*** The focus knob is used to vary the spread of the illumination beam based on the range and size of the area to be illuminated (Figure G-6).

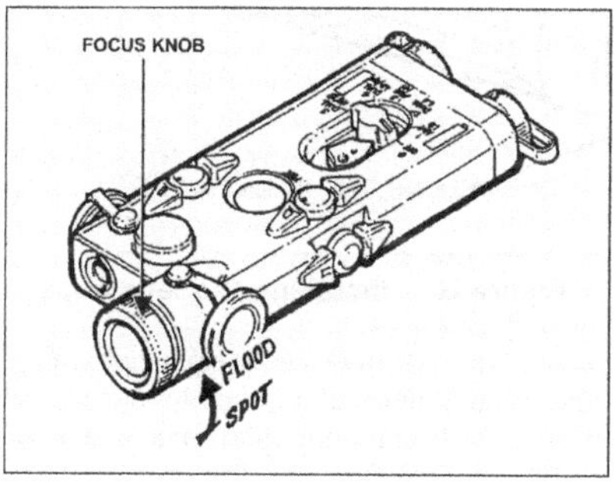

Figure G-6. Using the focus knob.

(7) ***Lens Caps.*** Several lens caps can be used with the AN/PEQ-2A for different purposes (Figure G-7).

(a) The *black lens cap* blocks the AN/PEQ-2A illuminator or aiming laser beam should the device be activated. To use the black lens cap, pull it from its stored location on the side of the unit and stretch it over the front of the focus knob or aiming beam so that it fits snugly in place.

(b) The *diffuser lens cap* enables the illuminator or aiming laser to emit in a 45-degree cone (10 feet at 10 feet). To use the diffuser lens cap, pull it over the front of the focus knob or aiming beam so that it fits snugly in place.

(c) The *neutral density lens cap* enables the AN/PEQ-2A illuminator or aiming laser to be operated in low power. To use the neutral density lens cap, pull it from its stored location on the side of the unit and stretch it over the front of the focus knob or aiming beam so that it fits snugly in place.

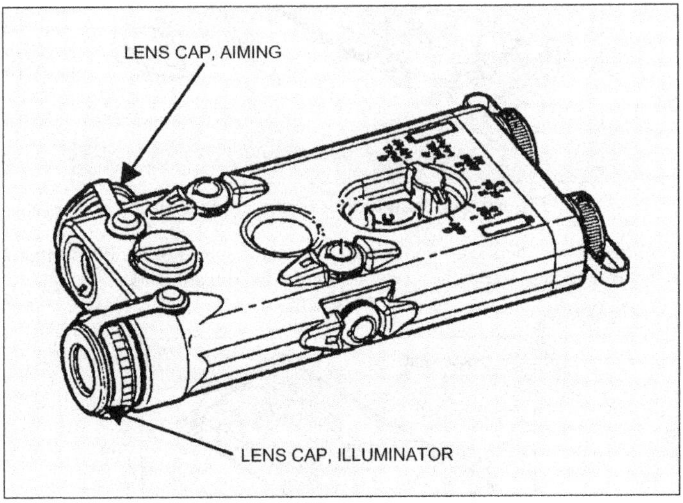

Figure G-7. Installing the lens caps.

(8) ***Boresight Adjusters.*** The AN/PEQ-2A is equipped with boresight adjusters for zeroing the aiming beam and illumination beam to the weapon (Figure G-8). The AN/PEQ-2A adjusters move the beam in true horizontal and vertical directions with the top adjusters used for elevation and the side adjusters for windage. When zeroing the AN/PEQ-2A, it is best to zero the aiming beam to the weapon and then align the illumination beam to the aiming beam.

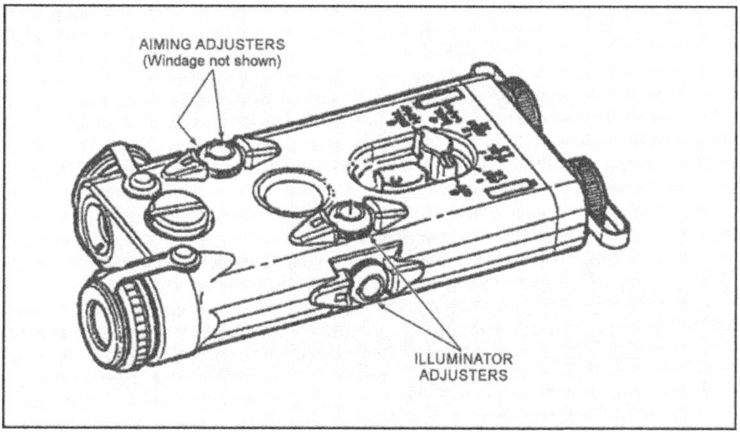

Figure G-8. Boresight adjusters for both aiming and illumination beams.

WARNING

Eye damage can occur if the laser is handled carelessly. The danger area is 15 meters in the training mode and 220 meters in the tactical mode.

b. **Mounting and Dismounting Procedures.** The AN/PEQ-2A can be mounted on various weapons or used in the hand-held mode to illuminate and designate targets.

(1) ***Mounting Procedures.*** Ensure that the mounting bracket is installed before proceeding.

(a) Secure the bracket adapter to the underside of the AN/PEQ-2A.

(b) Place the bracket adapter into the MK 19 bracket mounting groove, located in the center of the rail grabber.

(c) Tighten the bracket's lever screw into the bracket adapter hole.

(d) Plug the cable into the AN/PEQ-2A and run the cable to the front of the MK 19 bracket. The "cable switch, 12-inch, membrane" or the "cable switch, 20-inch, button" may be used for remote access.

(e) Once the cable is secure at the front of the MK 19 bracket (ensure the cable will not become damaged by the movement of the quadrant portion of the bracket) run it between the MK 19 and the bracket to the handles on the backplate assembly. If using the button cable, attach Velcro to the top of the night handle and secure excess cable to the handle. If using the membrane pad, run it the same way as with the button cable and attach the membrane to the inside of the right handle.

(2) ***Dismounting Procedures.*** Dismount the AN/PEQ-2A when the mission is complete, when it is needed on another weapon, or when it will be used in its hand-held mode.

(a) Untie the cable and disconnect the remote button from the backplate handles.

(b) Unplug the cable from the AN/PEQ-2A and place the cable back in the carrying case.

(c) Turn the AN/PEQ-2A mounting screw counterclockwise and remove the AN/PEQ-2A from the bracket adapter.

(d) Unscrew the bracket adapter from the MK 19 bracket rail and place the bracket adapter and the AN/PEQ-2A in its carrying case.

c. **Boresighting Procedures**. Boresighting aligns the sighting system to the bore of the weapon. The AN/PEM-1 Borelight is used for this procedure.

(1) Place the MK 19 in the ready to fire position 10 meters from the bore light offset zero target. Use the 10 meter distance gauge in the borelight kit to measure the distance. Ensure the target and weapon are level and stable (the bore sight zero will be off if not) before making any adjustments to the laser.

(2) Set the range on the MK 19 bracket to 500 meters.

(3) Mount the bore light and turn it on. While holding the MK 19 steady, zero the bore light. Follow the instructions in the AN/PEM-1 borelight operations manual for further details. The 40-mm MK 19 mandrel interface (NSN 3460-01-502-0575) is used with the 5.56-mm mandrel interface and has to be ordered separately.

(4) Make adjustments with the T&E mechanism until the bore light is on the bore light aiming point.

(5) Adjust the windage and elevation of the aiming laser until the laser is on the laser aiming point.

(6) Re-check the bore light aiming point and then the laser aiming point again.

(7) To zero the IR light source narrow the beam as much as possible and adjust its windage and elevation until the beam is 4 centimeters horizontally to the left of the laser aiming point while the aiming laser is still on the laser aiming point. The AN/PEQ-2A is now boresighted.

d. **Zeroing Procedures.** Zeroing the weapon aligns the line of sight of the AN/PEQ-2A with the trajectory of the round so that they intersect at 500 meters.

(1) Set the MK 19 mounting bracket to 500 meters by loosening the friction knob on the left side of the mount. Press and hold the friction knob in (towards the mount), then slide the mount assembly to the 500-meter increment. Release and tighten the friction knob.

(2) Identify a 500-meter target downrange and fire one round; note the impact of the round in relation to the target.

(3) Adjust the aiming light on the AN/PEQ-2A to the impact of the round by adjusting the elevation and windage adjusters.

(4) Manipulate the T&E mechanism so that the aiming light is center mass on the target. Repeat until the point of impact is the same as the point of aim.

(5) Adjust the illuminating beam's elevation and windage adjusters so that the aiming light is in the center of the illuminating beam.

G-2. THE AN/PAS-13B (V) 3 HEAVY WEAPON THERMAL SIGHT

The AN/PAS-13B (V) 3 HWTS is a silent, lightweight, compact, and durable battery-powered infrared imaging sensor that operates with low battery consumption (Figure G-9).

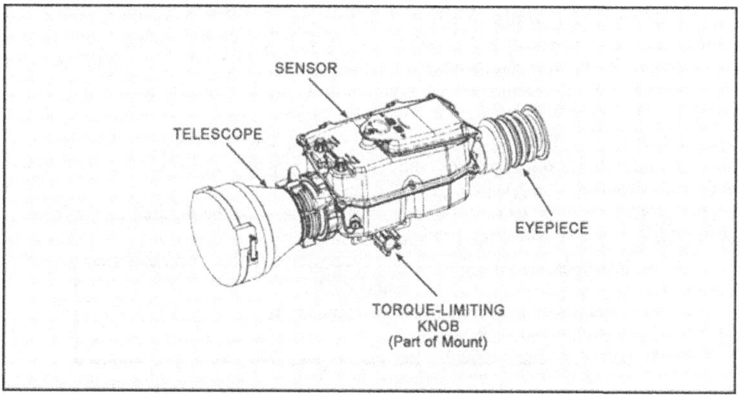

Figure G-9. AN/PAS-13B (V) 3 heavy weapon thermal sight.

a. **Components.** The HWTS is capable of target acquisition under conditions of limited visibility such as darkness, smoke, fog, dust, and haze. The HWTS operates effectively at night and can also be used in the daytime. Infrared light is received through the telescope, detected by an IR sensor, converted to digital data, processed, and then displayed for the user. Besides the carrying cases, the HWTS is composed of four major components: the telescope, the basic sensor, the eyepiece, and the mount.

(1) ***Telescope.*** The telescope receives IR light emitting from an intended target and its surroundings. The telescope magnifies and projects the IR light onto the basic sensor's scanner.

(2) ***Basic Sensor.*** The scanner reflects the IR light received from the telescope onto the detective assembly. The detective assembly senses the IR light and converts it to video. The sensor's electronics condition the video for display on the LED array. The LED array illuminates the IR image along with the reticle. The light from the LED array is reflected off the scanner to form an image at the eyepiece.

(3) ***Eyepiece.*** The eyepiece displays the thermal image and all system indicators on the cathode ray tube.

(4) ***Mount.*** A universal attachment interface between the HWTS and the MK 19 bracket (figure only shows the torque limiting knob).

b. **Modes of Operation.** The HWTS has three modes of operation: STANDBY, ON, and EMERGENCY.

(1) ***STANDBY Mode.*** When the system is first turned on, the HWTS begins a cool down period of approximately two minutes. After the cool down period, the HWTS

enters the STANDBY mode. During the STANDBY mode, power is not applied to the scanner or display in order to extend the life of the battery.

(2) ***ON Mode.*** When the HWTS is in the STANDBY mode and pressure is applied to the eyecup, the HWTS switches to the ON mode, and a switch engages to provide power to the scanner and display. After a three-second delay, the system is fully operational.

(3) ***EMERGENCY Mode.*** When switched to the EMERGENCY mode, the HWTS continuously applies power to the entire system. This allows the operator to bypass the three-second delay experienced when switching from the STANDBY to the ON mode. Since power is applied to the entire system while in the emergency mode, battery life is greatly reduced.

c. **Controls and Indicators.** Controls allow the gunner to configure the sight to the situation and individual preferences while the indicators display the current configuration (Figure G-10 and Table G-3).

(1) The CONTRAST CONTROL adjusts the contrast of the thermal image displayed on the raster. It has an automatic and a manual mode.

CAUTION

Ensure the brightness control switch is depressed before turning it on or off.

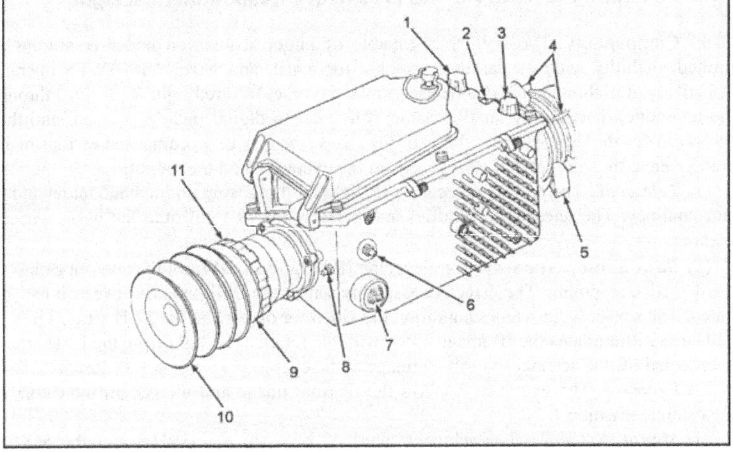

Figure G-10. Location of the HWTS controls and indicators.

ITEM	DESCRIPTION
1	Contrast Control
2	Emergency Switch Control
3	Brightness Control
4	Focus Ring
5	Field of View Ring
6	Zoom/Reticle Select Switch
7	Reticle Adjust Switch
8	Black/White Polarity Switch
9	Diopter Holding Device
10	Eyecup
11	Diopter Focus Ring

Table G-3. HWTS controls and Indicators.

(2) The EMERGENCY SWITCH CONTROL places the HTWS in emergency mode. The eyecup must be depressed in order to for the CRT to illuminate.

(3) The BRIGHTNESS CONTROL is a rotary switch with an off detent position (turned fully counterclockwise). The purpose is to turn the system on or off and adjust the brightness of the eyepiece display.

(4) The FOCUS RING adjusts the telescope focus from 20 meters to infinity. It requires a manual adjustment and affects both the wide and narrow fields of view.

(5) The FIELD OF VIEW RING (FOV) is located on the telescope. It has a wide and a narrow field of view. The wide FOV is for using low magnification during target detection, and the narrow FOV is for using high magnification during recognition and engagement.

(6) The RETICLE SELECT SWITCH selects one of the available reticles depending on the type of thermal sight (medium or heavy) and the weapon. It must be held for two seconds to enable reticle changes. After two seconds, release the switch to cycle to the next reticle. This control is disabled after ten seconds of inactivity.

(7) The RETICLE ADJUST SWITCH adjusts the reticle aiming features in azimuth and elevation. It is used during zeroing, and it must be held for two seconds to allow changes to be made. After two seconds, each press moves the reticle aiming features one increment. This control is also disabled after ten seconds of inactivity.

(8) The BLACK/WHITE POLARITY SWITCH selects the polarity of the thermal image displayed on the raster. The initial setting is "white hot." The polarity switch affects the appearance of the target.

(9) The DIOPTER HOLDING DEVICE allows the diopter to be adjusted when pulled back.

(10) The EYECUP controls the STANDBY/ON operating mode. When forward pressure is applied to the eyecup, the system is in the ON mode. When forward pressure is removed for more than 30 seconds, the system returns to the STANDBY mode.

(11) The DIOPTER FOCUS RING adjusts the focus of the raster and indicators to the operator's eye. It ranges from +2 to -6 diopters.

d. **Status Indicators and Raster** (Figure G-11). The status indicators and raster are visible when the user presses his eye against the eyecup. The indicators display the system status and configuration. The raster displays the thermal image, the selected reticle, and azimuth/elevation zeroing adjustment indicators.

(1) The status indicators are:

- NOT COOL (1): Displayed when the system is not cool enough for proper operation.
- WHT HOT/BLK HOT (2): Indicates the target polarity.
- EMER (3): Displayed when system is in emergency mode.
- LOW (4): Displayed when battery power is low (approximately fifteen minutes of useful power left).
- FIELD OF VIEW (7): Indicates when the wide field of view (WFOV) or the narrow field of view (NFOV) is in use
- ZOOM (8): Displayed when the zoom mode is selected
- RET ADJ (9): Indicates the reticle adjustment mode is selected
- RET SEL (10): Indicates the reticle select mode is selected

(2) The DISPLAY RASTER displays the thermal image with a superimposed reticle and indicators:

- RETICLE SELECT (5): Identifies the selected reticle/weapon.
- ELEVATION INDICATOR (6): Indicates the elevation zeroing adjustment of the reticle.
- AZIMUTH INDICATOR (11): Indicates the elevation zeroing adjustment of the reticle.
- RANGE SCALE (12): The range scale allows the operator to estimate the range to a target based on a known height or width. The numbers on the scale are the range in hundreds of meters (example: = 400 meters). The vertical lines on the scale reflect the height of a five-foot man at that range. The distance between two consecutive lines reflects the width of a 10-foot tank at the range specified by the vertical line on the left.
- CROSSHAIR/AIMING POINT (13): The vertical line at the aiming point reflects the height of a five-foot man at 800 meters. The horizontal line of the cross-hair reflects the width of a 10-foot tank at 800 meters.

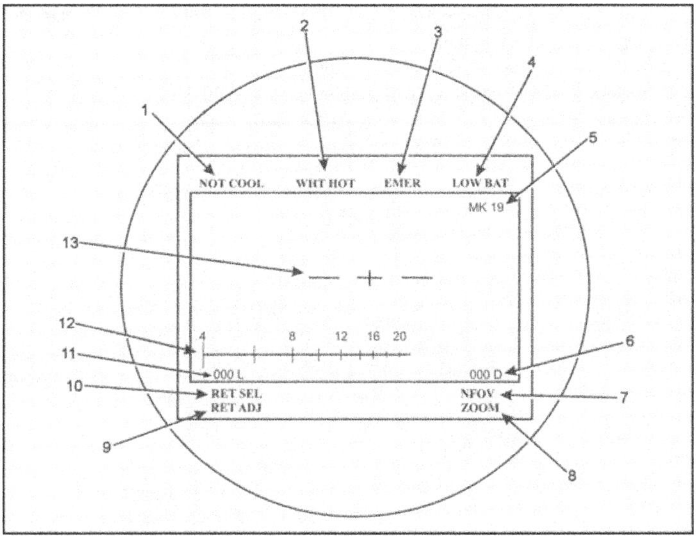

Figure G-11. Status indicators and raster.

e. **Mounting and Dismounting Procedures.** The HWTS is mounted on the MK 19 with a mounting bracket (Figure G-12). See TM 11-5855-312-10 for details.

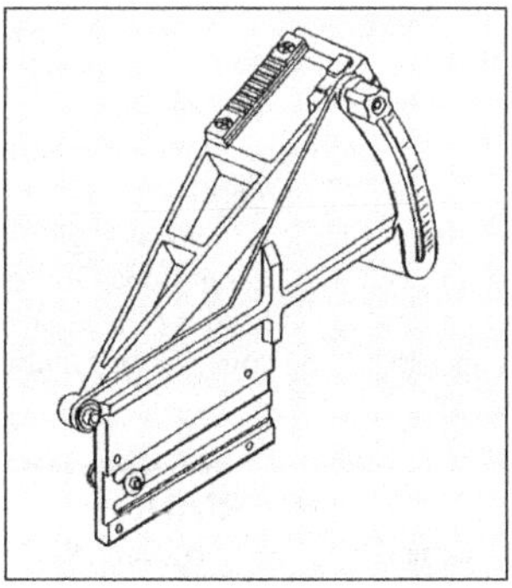

Figure G-12. MK 19 bracket.

(1) ***Mounting Procedures.*** Mounting the HWTS to the MK 19 is a two-step process (Figure G-13):

(a) Mount the bracket to the MK 19 by sliding the bracket into the dovetail slot until the locking pin engages the mount.

(b) Install the HWTS onto the bracket mount.

(c) Shake the HWTS to make sure the sight is secured properly.

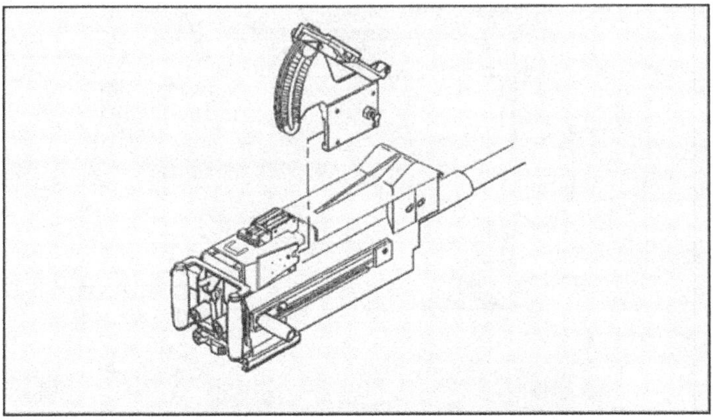

Figure G-13. Mounting the HWTS on MK 19.

(2) ***Dismounting Procedures.*** To dismount the HWTS:

(a) Take the HWTS off the mounting bracket rail and place the HWTS back in its carrying case.

(b) Pull out the locking pin and slide the bracket off the mount.

G-3. AN/TVS-5

The AN/TVS-5 is a portable, battery-operated electro-optical instrument used for observation and aimed fire of weapons at night (Figure G-14 and Table G-4). It amplifies reflected light such as moonlight, starlight, and sky glow so that the viewed scene becomes clearly visible to the operator. The sight does not emit visible or infrared light (except from the eyepiece) that can be detected by the enemy. By using this device, the gunner can observe the area and detect and engage any suitable target.

WARNING

Ensure the weapon is not loaded and is on S (SAFE) before installing the HWTS on the weapon. A loaded weapon may accidentally discharge causing severe injury or death.

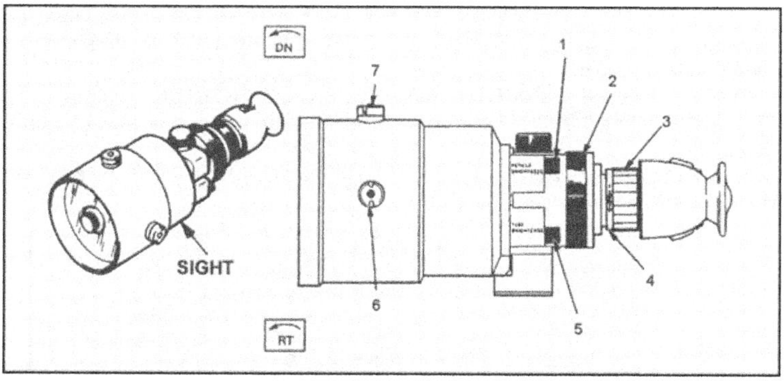

Figure G-14. Location of the AN/TVS-5 components.

ITEM	DESCRIPTION
1	ON-OFF/ Reticle Brightness
2	Objective Focus Ring
3	Diopter Focus Ring
4	Diopter Indicator
5	ON-OFF/ Tube Brightness
6	Reticle Azimuth Adjustment Acutator
7	Reticle Elevation Adjustment Acutator

Table G-4. AN/TVS-5 component names.

a. **Controls and Indicators.**Controls allow the gunner to configure the sight to the situation and individual preferences while indicators display the current configuration.

(1) The ON-OFF/TUBE BRIGHTNESS control applies power to the sight and controls the brightness of the image intensifier tube. It also enables the ON-OFF RETICLE BRIGHTNESS control to function.

(2) The ON-OFF RETICLE BRIGHTNESS control applies power to the reticle and controls the brightness of the reticle.

(3) The OBJECTIVE FOCUS RING adjusts the range focus from 25 meters to infinity.

(4) The DIOPTER FOCUS RING adjusts the focus of the eyepiece.

(5) The DIOPTER INDICATOR indicates the direction of rotation of the DIOPTER FOCUS RING for + and diopters.

(6) The RETICLE ELEVATION ADJUSTMENT ACTUATOR controls the reticle adjustment in the up and down direction. Each click of adjustment moves the strike of the round 1.0 inch at 100 meters.

(7) The RETICLE AZIMUTH ADJUSTMENT ACTUATOR controls reticle adjustment right or left. Each click or adjustment moves the strike of the round 1.0 inch at 100 meters.

b. **AN/TVS-5 Sight Reticle.** The MK 19 and the M2 machine guns share the same AN/TVS-5 sight reticle (Figure G-15).

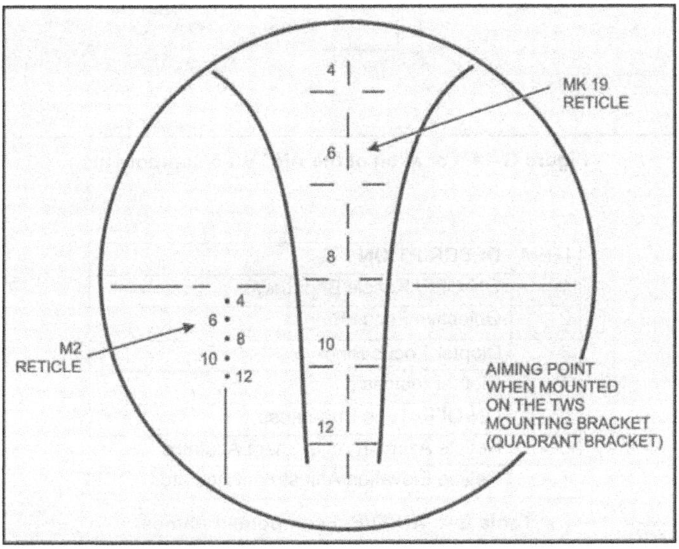

Figure G-15. AN/TVS-5 reticle for the MK 19 and M2 machine guns.

WARNING

1. **Eye guard:**
   - **To avoid recoil injury, attach the eye guard before using the sight.**
   - **When activated by pressure on the eye guard, the sight emits illumination the enemy can detect. Therefore, avoid touching the eye guard except when you wish to use the sight, and then press your eye area against it to activate it.**
2. **Batteries:**
   - **Dispose of batteries as directed. The contents of the mercury batteries cause extreme irritation to the eyes and to oral and nasal passages.**
   - **To prevent explosion, avoid disposing of batteries by burning.**
   - **Avoid short circulating batteries.**
   - **Avoid recharging batteries.**
   - **Remove batteries before storing sight.**
   - **Always replace both mercury batteries at the same time.**

c. **Mounting Procedures**. To mount the AN/TVS-5:

(1) Align the scribe line on the sight with the scribe line on the bracket.

(2) Place the sight in the grove of the bracket and tighten the lever screw clockwise.

(3) Secure the lever screw with lacing wire or tape to ensure the sight does not vibrate loose.

(4) Seat the device. After firing the initial burst, retighten the lever screw to ensure the sight is securely mounted. (If you are unable to fire at this time, lightly shake the sight to ensure the sight is mounted correctly.)

d. **Dismounting Procedures.** To dismount the AN/TVS-5:

(1) Remove the lacing wire or tape from the lever screw.

(2) Loosen the lever screw until the sight is free to be lifted off the bracket.

(3) Remove the batteries and place the sight in its carrying case.

e. **Boresighting.** The AN/TVS-5 cannot be boresighted when used with the MK 19. The sight cannot focus on a target closer than 25 meters, and the offset is too great at that distance to boresight accurately.

f. **Zeroing.** To zero the AN/TVS-5 to the MK 19, follow the procedures in TM 11-5855-214-10.

APPENDIX H

CREW DRILLS

The purpose of a crew drill is to build precision, speed, skill, and teamwork in checking equipment, placing the MK 19 into and out of action, and moving the MK 19 to a new position. The main goal is achieved by following prescribed procedures. Precision, speed, skill, and teamwork grow with practice. During the crew drill, all oral and visual signals are repeated. Once crewmembers can perform their duties well, they rotate to allow each member of the crew to learn the duties of others. The ground-mounted MK 19 has three crewmembers: the gunner, the assistant gunner, and the ammunition bearer. The vehicle-mounted MK 19 also has three crewmembers: the squad leader, the gunner, and the driver. Normally the crew is not required to carry the MK 19 and ammunition over long distances. A vehicle usually carries the crew, ammunition, and equipment to the firing position. The terrain and the enemy situation dictate how close the vehicle may be brought to the firing position.

H-1. CREW EQUIPMENT AND DRILL AREA

Dismounted crew drills are conducted with a gunner, assistant gunner, and ammunition bearer. For drills involving a vehicle, the squad leader, gunner, and driver are crewmembers. In addition to individual weapons and equipment, each crewmember for dismounted drills carries specific equipment used with the tripod-mounted MK 19 (Table H-1).

CREW MEMBER	SUGGESTED MINIMUM EQUIPMENT
Gunner	MK 19, binoculars, and compass
Assistant Gunner	M3 tripod, MK 64 gun cradle, with T&E mechanism attached
Ammunition Bearer	One box of ammunition

Table H-1. Equipment.

A crew drill is conducted in an area large enough to allow crewmembers to move freely with equipment and without interfering with the movements of other crewmembers.

> **WARNING**
> **Each soldier must wear all the protective equipment required by range safety regulations. This may include helmet, earplugs, and protective vest.**

H-2. FORMATION FOR CREW DRILL

To form the crew for crew drill, the leader commands FORM FOR CREW DRILL. Upon hearing the leader's command, the crew repeats the command and forms into a column, five paces apart, in the following order: assistant gunner, gunner, and ammunition bearer (Figure H-1).

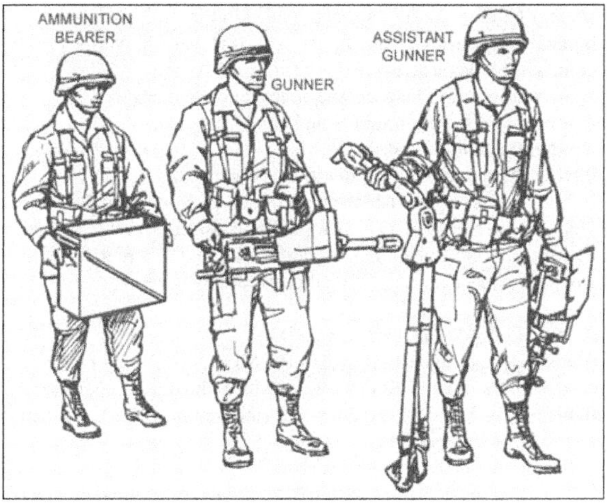

Figure H-1. Three-man MK 19 crew.

Each crewmember carries the following equipment in the specified manner:

- The assistant gunner carries the tripod, with the front leg positioned at a 60-degree angle and the trail legs collapsed, in his right hand; he carries the gun cradle, with the T&E mechanism attached, in his left hand.
- The gunner carries the MK 19, binoculars, and compass.
- The ammunition bearer carries one can of ammunition.

H-3. INSPECTION OF EQUIPMENT

The leader commands CHECK EQUIPMENT. All crewmembers repeat the command and inspect the components. The equipment is inspected and verified by the following crewmembers.

a. The assistant gunner inspects the M3 tripod, gun cradle, and T&E mechanism and ensures:

(1) The indexing levers and clamps on the front and trail legs work, and that the legs are in the "short (low)" position.

(2) The front leg is placed at a 60-degree angle and the trail legs are closely folded and the front leg clamp is tight.

(3) The sleeve lock latch and pintle lock release cam work.

(4) The pintle bushing is clean and has no burrs.

(5) The cradle pintle is clean, and all pins and bolts are seated.

(6) The T&E mechanism is centered.

b. The gunner inspects the MK 19 and ensures:

(1) The weapon is clean.

(2) The bolt is forward, the top cover is closed, and the weapon is on S (SAFE).

(3) The charger handles are forward and locked in the "up" position.

(4) The binocular lenses are clean and the focus mechanism works.

(5) The compass magnetic arrow and bezel ring work freely.

c. The ammunition bearer inspects the ammunition and ensures:

(1) The correct type of ammunition is on hand.

(2) The ammunition can is dent-free.

(3) All ogives are tight and the ammunition is clean.

d. The crewmembers report when the inspections are finished in the following manner:

(1) The ammunition bearer reports "Ammunition correct," or any defects.

(2) The gunner reports "Gun and ammunition correct," or any defects.

(3) The assistant gunner reports, "All correct," or any defects.

H-4. PLACING THE MK 19 INTO ACTION

The leader commands and signals GUN TO BE MOUNTED, and points to the position where the gun is to be mounted; FRONT, and points in the direction of fire; ACTION, and indicates with his fist the MK 19's direction in the chosen position. Crewmembers take the following steps to place the gun into operation.

a. The assistant gunner:

(1) Drags the tripod to the firing position by its front leg with his right hand, and carries the gun cradle with the T&E mechanism attached in his left hand.

(2) When he arrives at the firing position, aligns the front tripod leg for direction and, with one quick motion, snaps the trail legs apart. This action makes the sleeve lock-latch snap in place (Figure H-2).

(3) Stamps each leg into the ground and secures the gun cradle to the tripod by using the pintle lock release cam (Figure H-3).

(4) Secures the T&E mechanism to the tripod, by locking the traversing slide lock lever (Figure H-4).

(5) Helps the gunner mount and load the MK 19.

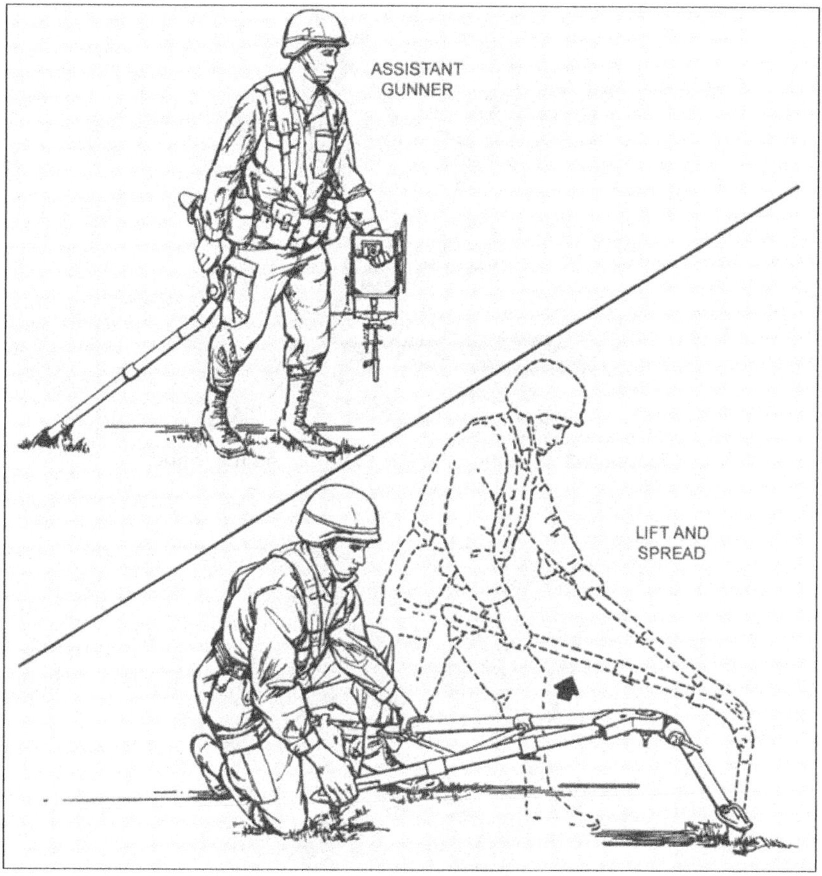

Figure H-2. Positioning the tripod.

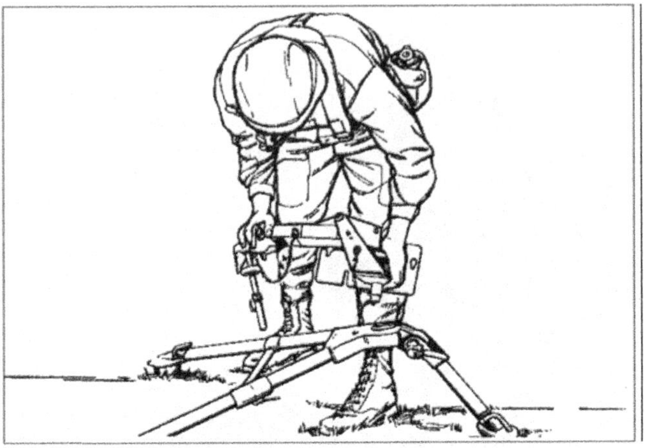

Figure H-3. Mounting the MK 19 mount on the tripod.

Figure H-4. Securing the T&E mechanism.

b. The gunner:

(1) Carries the MK 19 receiver to the tripod, with the barrel in his left hand, and the control grip in his right hand (Figure H-5).

(2) At the tripod, the assistant gunner grasps the barrel while the gunner holds the grips.

(3) The gunner and assistant gunner align the grooves on the receiver with the lugs in the gun cradle, and slide the receiver forward (Figure H-6).

(4) Secures the rear of the weapon by inserting the retaining pin through the cradle and sear assembly (Figure H-7).

Figure H-5. Gunner moving into position.

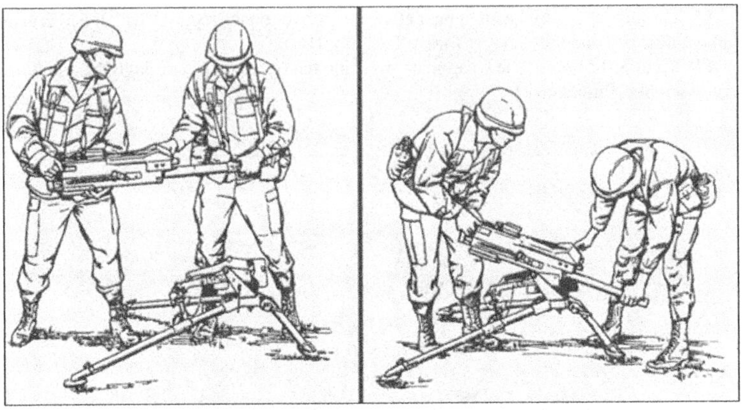

Figure H-6. Placing the MK 19 in the cradle.

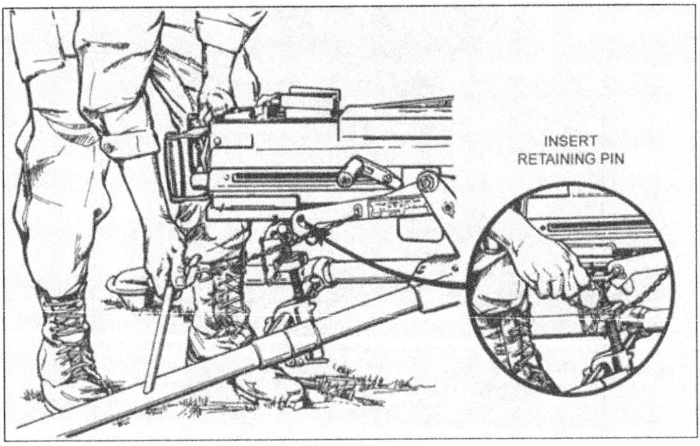

Figure H-7. Securing the MK 19 to the cradle.

c. The ammunition bearer:

(1) Carries the ammunition can to the MK 19 position (Figure H-8).

(2) Takes the lid off and ensures the ogive ends of the rounds are pointed downrange (Figure H-9).

(3) Gives the can to the assistant gunner.

(4) Returns to the ammunition point and stands by to do whatever is needed to complete the mission.

Figure H-8. Ammunition bearer with ammunition can.

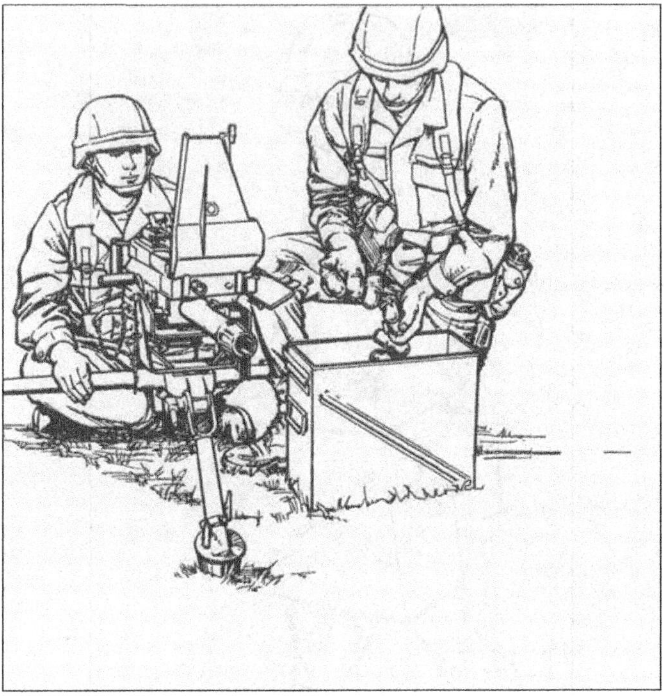

Figure H-9. Alignment of ammunition belt.

d. The gunner opens the top cover. The assistant gunner inserts the first round into the feeder (female link first), pushes the first round across the primary feed pawl (Figure H-10), and moves the feed slide assembly to the left by pushing the secondary drive lever to the right (Figure H-11).

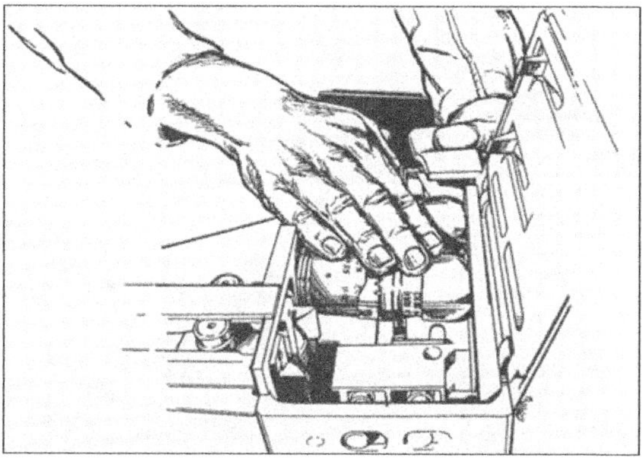

Figure H-10. Placement of the round across the primary feed pawl.

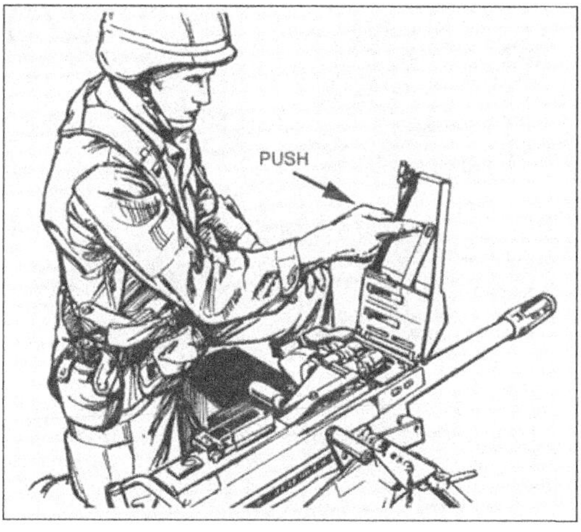

Figure H-11. Feed slide assembly moved to the left.

e. The gunner closes the top cover and charges the weapon (Figure H-12). The gunner presses the trigger, which places the lead round in the bolt extractor. He charges the weapon a second time, forcing the lead round from the bolt extractors into the bolt fingers. The weapon is ready to fire. When the assistant gunner raises the rear sight, the gunner calls "Up" (Figure H-13).

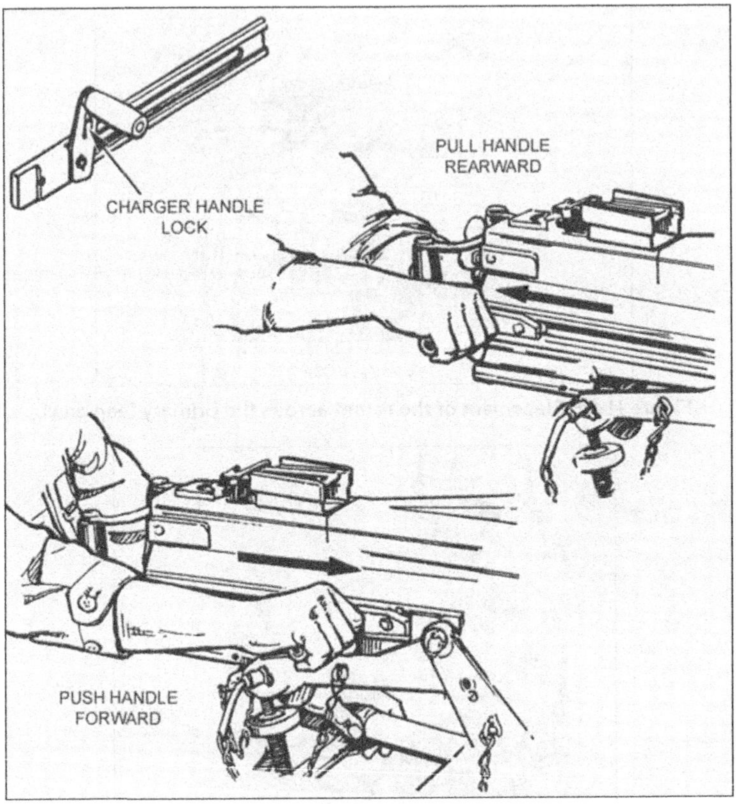

Figure H-12. Gunner charging the MK 19.

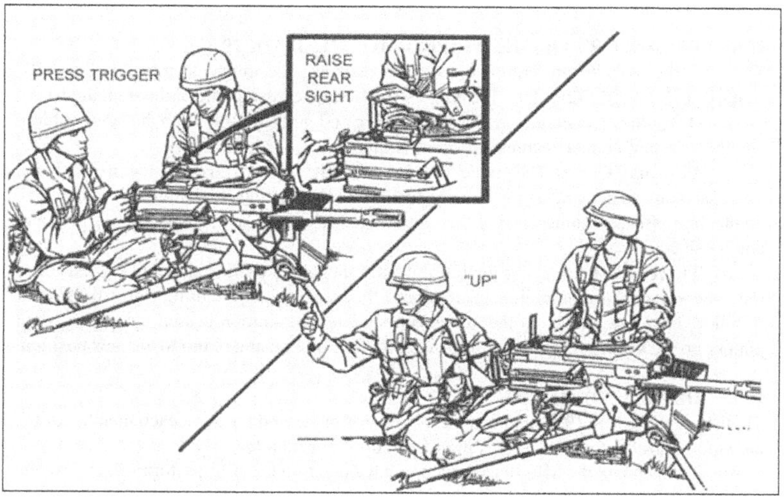

Figure H-13. Raising the rear sight.

H-5. TAKING THE MK 19 OUT OF ACTION

To take the gun out of action, the leader commands OUT OF ACTION. Each crewmember accomplishes the following in sequence.

a. The gunner raises the top cover of the MK 19. The assistant gunner releases the ammunition belt from the feed tray and places it in the ammunition can. The gunner takes the round or spent cartridge case from the bolt fingers, places it in the ammunition can, and closes the top cover.

b. The assistant gunner releases the MK 19 from the gun cradle. The gunner lifts the MK 19 off the mount and moves to a position 10 paces behind the tripod. He places the MK 19 on the ground in front of him with the flash suppressor to the left and with the sights on the top.

c. Once the gunner has moved away from the tripod, the assistant gunner:

(1) Releases the T&E mechanism and takes off the gun cradle/ T&E mechanism.

(2) Releases the tripod sleeve lock latch.

(3) Grasps the front leg and rotates the tripod to a vertical position.

(4) Releases the sleeve latch, and collapses the trail legs.

(5) Loosens the front leg clamp, folds down the front leg to a 60-degree angle, and tightens the clamp.

(6) Returns, with the tripod and gun cradle, to a position five paces in front of the gunner.

d. The ammunition bearer comes forward and gets the ammunition can(s) or any other loose equipment, and returns to a position five paces behind the gunner.

e. The gunner announces "Up."

H-6. TRANSPORT OF THE TRIPOD-MOUNTED MK 19

To move the MK 19 and equipment a short distance, the gunner commands SECURE EQUIPMENT, FOLLOW ME. The MK 19 can be moved using one of three methods:

a. **Dragging.** The gunner grasps the front leg and pulls the MK 19 to the new position, and the assistant gunner secures any loose equipment.

b. **Two-man Carry.** The gunner stands behind the tripod with a trail leg in each hand. The assistant gunner grasps the front leg and avoids touching the barrel if it is hot. The gunner and assistant gunner carry the mounted MK 19. The assistant gunner returns to secure any loose equipment.

c. **Three-man Carry.** The gunner and assistant gunner each holds a trail leg just above the traverse bar. The ammunition bearer grasps the front leg and avoids touching the barrel if it is hot. The crew moves to the new position. The ammunition bearer and the assistant gunner go back and get the ammunition or loose equipment, and return to the new position.

H-7. DISMOUNT THE MK 19

To dismount the MK 19 from the vehicle-mounted or stowed position, each member of the crew performs the following steps in order within four minutes.

a. To dismount the MK 19 from the mounted position the crew performs the following actions.

(1) The squad leader:

(a) Dismounts the vehicle with binoculars and compass.

(b) Obtains the M3 tripod from the stow position.

(c) Takes the M64 gun cradle and T&E mechanism from the gunner and moves to the ground firing position.

(d) At the firing position, the gunner aligns the front tripod leg for direction and snaps the trail legs apart.

(e) Stamps the tripod legs into the ground.

(f) Zero's the T & E mechanism, if it is not zeroed.

(g) Secures the gun cradle to the M3 tripod and the T&E mechanism to the traverse bar and gun cradle.

(h) Assists the gunner in mounting the MK 19.

(2) The gunner:

(a) Raises the top cover, releases the ammunition belt from the feed tray, and places it in the ammunition can.

(b) Takes the round or spent cartridge case from the bolt fingers, places it in the ammunition can, and closes the top cover.

(c) Lifts the MK 19 from the gun cradle after the driver releases it, and hands it to the driver.

(d) Releases the T&E mechanism from the pedestal with the train lock handle and removes the empty case catch bag assembly.

(e) Removes the gun cradle with the T&E mechanism attached and hands it to the squad leader.

(f) Takes the MK 19 from the driver and moves to the firing position.

(g) With the squad leader's assistance, the gunner secures the MK 19 to the gun cradle.

(3) The driver:

(a) Dismounts the vehicle and assists the gunner in clearing the weapon.

(b) Removes the ammunition can and releases the MK 19 and takes it from the gunner.

(c) Gives the MK 19 to the gunner when the gunner is ready.

(d) Stows the empty case catch bag assembly and takes the ammunition can to the firing position, while also removing the lid and ensuring the ogive ends of the rounds are pointed downrange.

(e) Returns to the vehicle and monitors the radio and/or takes more ammunition to the firing point, and maintains rear security.

(4) The squad leader and the gunner load the weapon (as in a crew drill), and place the MK 19 into action.

b. To dismount the MK 19 from the stowed position the crew performs the following actions.

(1) The squad leader:

(a) Secures the binoculars and compass.

(b) Obtains the M3 tripod, the M64 gun cradle, and the T&E mechanism from their stowed position.

(c) Moves to the ground firing position where he aligns the front tripod leg for direction and snaps the trail legs apart.

(d) Stamps the tripod legs into the ground.

(e) If required, zeros' the T&E mechanism to the traverse bar and gun cradle.

(f) Assists the gunner in mounting the MK 19.

(2) The gunner:

(a) Dismounts the vehicle and obtains the MK 19 from its stowed position.

(b) Moves to the firing position with the MK 19.

(c) Secures the MK 19 into the gun cradle with the squad leader's assistance.

(3) The driver:

(a) Dismounts the vehicle and moves to the firing position with a can of ammunition.

(b) Assists in obtaining both the M3 tripod and the MK 19 from the stowed position.

(c) Removes the lid from the ammunition can and ensures the ogive ends of the rounds are pointed downrange.

(d) Returns to the vehicle, monitors the radio and/or takes more ammunition to the firing position, and maintains rear security.

(4) The squad leader and the gunner load the weapon as in the crew drill and place the MK 19 into action.

H-8. REMOUNT THE MK 19

The vehicle crew must be able to mount the MK 19 from the ground-mount to a vehicle-mount or stowed position within four minutes.

a. To mount the MK 19 from the ground to the vehicle the crew performs the following actions.

(1) The squad leader assists the gunner in clearing the weapon by releasing the ammunition belt from the feed tray, and placing it in the ammunition can. He then:

(a) Removes the ammunition can and places it on the ground.

(b) Releases the MK 19 from the gun cradle.

(c) Releases the T&E mechanism from the tripod traverse rod and takes off the gun cradle and T&E mechanism, after the gunner removes the MK 19.

(d) Grasps the front leg of the tripod and rotates the tripod to vertical position, releases the sleeve latch, and collapses the trail legs.

(e) Loosens the front leg clamp, folds down the front leg, and tightens the clamp.

(f) Returns to the vehicle, gives the gun cradle/T&E mechanism to the gunner, and stows the M3 tripod.

(g) Stows the binoculars and compass in the vehicle and resumes his duties.

(2) The gunner raises the top cover of the MK 19 and takes the round or spent cartridge case from the bolt fingers, and places it in the ammunition can, and closes the top cover. He then:

(a) Lifts the MK 19 off the gun cradle when the squad leader releases it and moves to the vehicle.

(b) Gives the MK 19 to the driver and mounts the vehicle.

(c) Takes the gun cradle and T&E mechanism from the squad leader and mounts the gun cradle to the pedestal.

(d) Zeros' the T&E mechanism if needed and reattaches it to the gun cradle.

(e) Attaches the T&E mechanism to the pedestal and clamps it with the train lock handle.

(f) Takes the MK 19 from the driver and installs it on the gun cradle.

(g) Attaches the travel lock assembly if movement is imminent.

(h) Takes the empty case catch bag assembly from the driver and attaches it to the gun cradle.

(3) The driver receives the MK 19 from the gunner and gives it to him when he is ready. He then:

(a) Assists the gunner in installing the MK 19 in the gun cradle.

(b) Moves to the firing position, retrieves the ammunition can and any loose equipment, and returns to the vehicle.

(c) Places the ammunition can on top of the vehicle, with the ogive ends of the rounds pointed down range.

(d) Hands the empty case catch bag assembly to the gunner.

(4) The squad leader and the gunner load the weapon (as in the crew drill) and place the MK 19 into action.

b. To dismount the MK 19 from the ground mount to a stowed position, the crew performs the following actions.

(1) The squad leader:

(a) Assists the gunner in clearing the weapon by releasing the ammunition belt from the feed tray, and places it in the ammunition can.

(b) Removes the ammunition can and places it on the ground. He then releases the MK 19 from the gun cradle.

(c) Releases the T&E mechanism from the tripod traverse rod and takes off the gun cradle/T&E mechanism after the gunner removes the MK 19.

(d) Grasps the front leg of the tripod and rotates the tripod to a vertical position, releases the sleeve latch, collapses the trail legs, loosens the front leg clamp, folds down the front leg, and tightens the clamp.

(e) Returns to the vehicle and stows the M3 tripod, T&E mechanism, binoculars, and compass.

(f) Resumes his duties.

(2) The gunner:

(a) Raises the top cover of the MK 19 and takes the round or spent cartridge case from the bolt fingers.

(b) Places the round in the ammunition can, and closes the top cover.

(c) Lifts the MK 19 off the gun cradle when the squad leader releases it and moves to the vehicle.

(d) Stows the MK 19 in its stowage position.

(3) The driver:

(a) Moves to the firing position.

(b) Collects the ammunition can and any loose equipment.

(c) Stows the ammunition can and assists the squad leader and gunner in stowing any of their equipment.

H-9. CONVERT THE M966 HMMWV TOW MISSILE CARRIER FROM MK 19 MODE TO TOW 2 MODE

To convert the M966 weapon system from the MK 19 mode to the TOW 2 mode, the crew performs the following actions.

a. The squad leader:

(1) Places the traversing unit in the traveling position.

(2) Obtains the nightsight and collimation cases from the stowed position and places them on top of the vehicle near the gunner.

(3) Stows the gun cradle/ammo bracket, T&E mechanism, and empty case catch bag assembly.

(4) Unstraps the missile guidance system (MGS), unlatches its cover, and stows the MGS in the commander's footwell.

(5) Connects the J-1 cable and the battery power conditioner cable to the MGS and MGS battery power conditioner.

(6) Connects the nightsight power conditioner cable to the TOW vehicle power cable (TVPC), uncoils the cable and hands it up to the gunner.

(7) Ensures that the power cable is connected to the TVPC, and that the TVPC switch is on.

(8) Assists the gunner in boresighting the system.

b. The gunner :

(1) Raises the top cover and releases the ammunition belt from the feed tray and places it in the ammunition can.

(2) He takes the round or spent cartridge case from the bolt fingers, places it in the ammunition can, and closes the top cover.

(3) The gunner releases the MK 19 from the gun cradle, lifts the MK 19 out of the gun cradle, and gives it to the driver.

(4) The gunner releases the T&E mechanism from the pedestal with the train lock handle.

(5) Removes the empty case catch bag assembly and gives it to the squad leader. The gunner moves the ammo can and gives it to the driver.

(6) Removes the gun cradle with the T&E mechanism attached and hands it to the squad leader.

(7) Opens the pedestal mount cover and secures it. Lifts the traversing unit with adapter from the stowage base and secures the traversing unit to the pedestal mount with the coupling clamp.

(8) Receives the launch tube from the driver and secures it to the traversing unit.

(9) Receives the optical sight from the driver and installs it on the traversing unit.

(10) Opens the nightsight case, removes nightsight device and installs it on the optical sight and hands the case to the driver.

(11) Takes the TVPC cable from the squad leader and installs the nightsight device.

(12) Opens the collimation case, removes the collimator, attaches to nightsight, and connects collimator power cable.

(13) Assists the squad leader in boresighting.

(14) Collimates sights, removes collimator and power cable, stores collimator and cable in case, and hands case to the driver.

(15) Receives the missile from the driver and loads it in the system.

c. The driver:

(1) Dismounts the vehicle and enters the right rear passenger door and then unlocks and opens the cargo hatch.

(2) Stows the MK 19 when received from the gunner.

(3) Stows the ammunition can when received from the gunner.

(4) Places the optical sight on top of the vehicle for the gunner.

(5) Unstraps a missile, inspects the missile, and gives the missile to the gunner.

(6) Ensures every item in the cargo hatch is secure, closes the cargo hatch, exits the vehicle through the right rear passenger door and takes his place as driver.

H-10. CONVERT THE M966 HMMWV TOW MISSILE CARRIER FROM TOW 2 MODE TO MK 19 MODE

This exercise requires the crew to accomplish its task within three minutes. To convert the M966 weapon system from the TOW 2 mode to the MK 19 mode and put the TOW 2 into its stowed position, the crew performs the following actions.

a. The squad leader:

(1) Disconnects the J-1 cable from the MGS, coils the cable and stows in the traversing unit.

(2) Disconnects the battery power conditioner cable from the MGS battery power cable, and winds it on the stowage handle.

(3) Disconnects the MGS battery cable, recovers the MGS cover, stows the cable in the MGS, replaces the cover on the MGS, and straps the MGS down.

(4) Disconnects the nightsight power cable from the TVPC, and winds and stows the cable.

(5) Receives the nightsight case from the gunner and stows it in the vehicle.

(6) Gives the gunner the gun cradle/ammo bracket, T&E mechanism, and empty case catch bag assembly.

b. The gunner:

(1) Arms the missile (if armed), unlocks the bridge clamp, removes the missile and hands it to the driver.

(2) Disconnects the TVPC power cable from the night sight and drops down the hatch.

(3) Disconnects the nightsight and places it in the case, and hands the case to the squad leader.

(4) Disconnects the optical sight and hands it to the driver.

(5) Disconnects the launch tube, and hands it to driver.

(6) Unclasps the traversing unit, lifts the traversing unit out of the pedestal mount, and takes it through the gunners hatch.

(7) Secures it to the TU stowage base, and closes the pedestal mount access cover.

(8) Takes the gun cradle/ammo bracket and T&E mechanism from the squad leader.

(9) Installs the cradle/ammo bracket on the heavy interdiction missile system (HIMS) mount.

(10) Ensures the T&E mechanism is zeroed, installs the T&E mechanism to the gun cradle and the pedestal mount on the HIMS.

(11) Takes the MK 19 from the driver and installs it on the gun cradle.

(12) He installs the empty case catch bag assembly on the gun cradle.

(13) Receives ammunition can from the driver, while ensuring that the ogive ends are pointing down range, and loads the MK 19.

c. The driver:

(1) Dismounts the vehicle and enters the right rear passenger door, and unlocks and opens the cargo hatch.

(2) Takes the missile from the gunner and stows it in the missile rack.

(3) Unstraps the nightsight case and gives it to the gunner.

(4) Takes the optical sight from the gunner and stows it in the vehicle.

(5) Takes the launch tube from the gunner and stows it in the vehicle.

(6) Unstraps the MK 19 from its stowage, and gives it to the gunner.

(7) Unstraps one can of ammunition and places it on top of the vehicle near the gunner.

(8) Replaces the forward handling ring and dust cover on the missile.

(9) Ensures that every item in the cargo hatch is secure and closes the cargo hatch.

(10) Exits the vehicle through the right rear passenger door and takes his place as the driver.

H-11. ENGAGE TARGETS WITH THE MK 19

The MK 19 squad must be able to successfully engage a target while occupying a firing position. This crew drill is based on the procedures found in Section III of Chapter 5, Application of Fire. The purpose of this drill is to exercise proper fire techniques to successfully engage a target while not hitting friendly troops or attempting to hit a target out of range. The crew occupies a dismounted position when the squad leader announces, "Fire mission."

a. The squad leader then performs the following tasks:

(1) Announces direction using one of the following methods: orally (example:"Right front," or "Left flank"), by a reference point (example:"TRP 1," "Lone tree."), fires tracer from a weapon or another MK 19, or by laying the gun on the target himself.

(2) Announces target description "Troops," "Tank," or if more than one type of target, designated target, "Lead truck," "Right building, far end."

(3) Announces target range, for example, "Four five zero," "One one hundred."

(4) Takes the following corrective actions, as required. If the gunner announces, "Cannot identify," the squad leader attempts to identify and gives further instructions. If the gunner announces "Lost," the squad leader gives additional target location. If the gunner announces "Friendly, the squad leader gives new target description or gives end of mission.

(5) Announces the method of engagement by announcing the class and rate of fire.

(6) Announces method of control.

(7) Issues subsequent fire commands as needed.

(8) Terminates the alert: CEASE FIRE, END OF MISSION.

b. The gunner performs the following tasks:

(1) Attempts to acquire target. If he cannot, the gunner announces "Cannot identify," "Lost," or "Friendly."

(2) When identified, announces "Identified," or "Up."

(3) Makes an adjustment to the lay of the gun as the squad leader gives new data.

c. The driver performs the following tasks:

(1) Provides security for the vehicle or position.

(2) Prepares to get more ammunition to the gun as required.

(3) Prepares to move if the squad leader announces "Cease fire," or "End of mission."

APPENDIX I

GUNNERY EXERCISES

There are seven exercises used to train MK 19 gunners and crews. The first four exercises are individual gunnery firing exercises and allow the gun crew to practice firing prior to the collective gunnery exercises. Commanders dictate the level of MOPP used during MK 19 gunnery.

These seven tables are used for single and multiple gun-crew practice and can be conducted on any MK 19 range. Each table has columns for the task, condition, standard, ammunition required, time required, and a column to record a satisfactory or unsatisfactory (SAT Y/N) score for the task.

I-1. INDIVIDUAL GUNNERY

These first four tables are used for gunnery practice. Each single gun exercise requires the gun crew to mount and dismount the MK 19, and engage point and area targets at various ranges. The two dismounted day exercises require the gun crew to prepare a range card and engage targets using data from the range card (Tables I-1 and Table I-2). The third exercise (Table I-3) is a mounted combat exercise and the fourth exercise (Table I-4) is a night fire exercise.

a. **Instructional Fire Exercise** (Table I-1**)**. The instructional firing exercise gives the crew practice firing on a range.

ACTION	CONDITIONS	STANDARDS	AMMO	TIME	SAT Y/N
1. Mount the MK 19 on the M3 tripod.	Given a MK 19, M3 tripod, T&E mechanism, MK 64 gun cradles, and a firing position.	Gun crew must be able to mount the MK 19 in its tripod mode within 1 min.	0	1 min	
2. Prepare a range card.	Given a MK 19 mounted on a M3 tripod, and a firing position with a designated sector of fire and targets.	Gunner must prepare a completed range card within 15 min.	0	15 min	
3. Zero the MK 19 while using a 400-meter BMP target.	Given a tripod-mounted MK 19, 400-meter BMP target, selected firing position, and 4 rds of 40-mm TP.	Gunner must impact on the target with at least 2 of the 4 rds on the BMP target at 400 meters.	4 rds	N/A	
4. Engage a BMP target at 1,100 meters while using range card data (point target).	Given a tripod-mounted MK 19, completed range card, BMP target at 1,100 meters, selected firing position, and 8 rds of 40-mm TP.	Gunner must impact on the 1,100 meter BMP target with at least 2 out of 8 rounds.	8 rds	N/A	
5. Engage a BMP target at 1,500 meters while using range card data.	Given a tripod-mounted MK 19, completed range card, BMP target at 1,500 meters, selected firing position, and 8 rds of 40-mm TP.	Gunner must impact on the 1,500 meter BMP target area with at least 1 out of 8 rds.	8 rds	N/A	
6. Engage dismounted troops at 600 meters while using range card data (area target).	Given a tripod-mounted MK 19, completed range card, dismounted troop targets at 600 meters, selected firing position, and 6 rds of 40-mm TP.	Gunner must impact at least 4 out of 6 rds within 5 meters of a silhouette at a range of 600 meters.	6 rds	N/A	
7. Dismount the MK 19.	Given a MK 19, M3 tripod, and a selected firing position.	Gun crew must dismount the MK 19 from the M3 tripod within 1 min.	0	1 min	
			Total: 26 rds		

Exercise Scoring

Fully Trained (T) = Satisfactory on 7 of 7 tasks
Need practice (P) = Satisfactory on 5 or 6 tasks
Untrained (U) = Unsatisfactory on 4 or fewer tasks

Table I-1. Instructional fire exercise.

During scoring procedures, the grader positions himself so that he can observe both the gunner and the target. Once the exercise commences, he:

- Times the tasks having a time standard.
- Observes and informs the gunner of the strike of the rounds.
- Records the results of each task in the right hand column with a "Y" for a satisfactory completion of the task or an "N" for an unsatisfactory completion.
- Sums the results and assigns an overall score in accordance with the scores in the bottom box of the figure above.

b. **Dismounted Range Card Exercise** (Table I-2). The dismounted range card exercise is held after instructional fire and gives the gunner and crew a chance to test their skills against a time-and-hit exercise. It is held as often as the commander feels necessary to maintain crew skills.

(1) Time the gunner and crew while they place the tripod-mounted MK 19 into operation in a selected location.

(2) Time the gunner while he completes a range card with the ranges of all targets within his sector of fire.

(3) Have the gunner suppress or kill single or multiple targets in his sector of fire while using his range card data.

(4) Time the gunner and crew while they take the MK 19 out of operation.

ACTION	CONDITIONS	STANDARDS	AMMO	TIME	SAT Y/N
1. Mount the MK 19 on an M3 tripod.	Given a MK 19, M3 tripod, M64 gun mount, T&E mechanism, and a selected firing position.	Gun crew must mount the MK 19 on the M3 tripod within 1 min.	0	1 min	
2. Prepare a range card.	Given a MK 19 mounted on a M3 tripod, a firing position with an assigned sector of fire, and designated targets.	Gunner must prepare a completed range card with all targets in his sector of fire within 15 min.	0	15 min	
3. Zero the MK 19 while mounted on the M3 tripod at a 400-meter BMP.	Given a tripod-mounted MK 19, selected firing position, 400-meter BMP target, and 4 rds of 40-mm TP.	Gunner must impact at least 2 out of 4 rds on the 400-meter BMP target.	4 rds	N/A	
4. Engage a 600-meter and tripod-mounted BMP target while using range card data (point target).	Given a tripod-mounted MK 19, selected firing position, 600-meter BMP target, and 4 rds of 40-mm TP.	Gunner must impact at least 3 out of 4 rds on the 600-meter BMP target within 1 min.	4rds	1 min	

Table I-2. Dismounted range card exercise.

ACTION	CONDITIONS	STANDARDS	AMMO	TIME	SAT Y/N
5. Engage a 1,100-meter BMP target while using range card data.	Given a tripod-mounted MK 19, selected firing position, BMP target at 1,100 meters, completed range card, and 8 rds of 40-mm TP.	Gunner must impact on BMP target at 1,100 meters with at least 2 out of 8 rds within 2 min.	8 rds	2 min	
6. Engage a 600-meter (area) troop target while using the range card data.	Given a tripod-mounted MK 19, 6 rds of 40-mm TP, selected firing position, and a troop target at 600 meters.	Gunner must impact at least 4 out of 6 rds within 5 meters of a troop silhouette at a range of 600.	6 rds	1.5 min	
7. Engage a BMP target at 1,500 meters while using the range card data.	Given a tripod-mounted MK 19, 8 rds of 40-mm TP, selected firing position, BMP target at 1,500 meters, and a completed range card.	Gunner must impact on the 1,500 meter BMP target with at least 4 out of 8 rds within 2 min.	8 rds	2 min	
8. Engage a BMP target at 1,100 meters and dismounted troops at 600 meters while using range card data.	Given a tripod-mounted MK 19, 12 rds of 40-mm TP, BMP target at 1,100 meters, dismounted troop target at 600 meters, and a completed range card.	Gunner must impact at least 2 rds on the BMP target at 1,100 meters, then shift to a dismounted troop target and place at least 2 rds within 5 meters of a troop silhouette at a range of 600 meters.	12 rds	3 min	
9. Dismount the MK 19 from the M3 tripod.	Given a mounted MK 19.	Gun crew must take the MK 19 out of operation from the tripod mount within 1 min.	0	1 min	
			Total: 42 rds		
Exercise Scoring **Fully Trained (T)** = Satisfactory on 9 of 9 tasks **Need Practice (P)** = Satisfactory on 7 or 8 tasks **Untrained (U)** = Unsatisfactory on 6 or fewer tasks					

Table I-2. Dismounted range card exercise (continued).

During scoring procedures, the grader positions himself so that he can observe both the gunner and the target. Once the exercise commences, he:

- Times the tasks having a time standard.
- Observes and informs the gunner of the strike of the rounds.
- Records the results of each task in the right hand column with a "Y" for a satisfactory completion of the task or an "N" for an unsatisfactory completion.
- Sums the results and assigns an overall score in accordance with the scores in the bottom box of the figure above.

c. **Mounted Combat Exercise.** The mounted combat exercise gives the gunner and the crew practice fighting from a vehicle. It is held as often as the commander feels necessary to maintain crew skills (Table I-3).

(1) Time the gun crew while they place the MK 19 into operation on its organic carrier. Carriers may be HMMWVs, 2 1/2-ton trucks, M151-series utility trucks, or M113-series APCs.

(2) Using the time-and-hit standards, give the gunner and crew a series of single and multiple targets to suppress or kill in their sector of fire.

(3) Upon completion of mounted firing, time the gunner and crew while they take the MK 19 out of operation.

ACTION	CONDITIONS	STANDARDS	AMMO	TIME	SAT Y/N
1. Mount the MK 19 on its organic carrier.	Given a MK 19, organic carrier, and a selected firing position.	Gun crew must mount the MK 19 on its organic carrier within 2 min.	0	2 min	
2. Engage a BMP target at 400 meters while mounted.	Given a MK 19, organic carrier, 4 rds of 40-mm TP, and a selected firing position.	Gunner must impact on the BMP target with at least 3 out of 4 rds at 400 meters within 1 min.	4 rds	1 min	
3. Engage a BMP target at 800 meters while mounted.	Given a MK 19, organic carrier, 8 rds of 40-mm TP, and a selected firing position.	Gunner must impact a BMP target at 800 meters with at least 3 out of 8 rds within 1.5 min.	8 rds	1.5 min	
4. Engage a 600-meter (area) target.	Given a MK 19, organic carrier, 6 rds of 40-mm TP, and a selected firing position.	Gunner must impact at least 4 out of 6 rds within 5 meters of a troop silhouette at a range of 600 meters.	6 rds	1.5 min	
5. Engage a BMP target at 1,100 meters.	Given a MK 19, organic carrier, 8 rds of 40-mm TP, and a selected firing position.	Gunner must impact the BMP target area at 1,100 meters with at least 2 out of 8 rds within 2 min.	8 rds	2 min	
6. Dismount the MK 19 from its organic carrier.	Given a MK 19 and an organic carrier.	Gun crew must dismount the MK 19 from its organic carrier within 2 min.	0	2 min	
			Total: 26 rds		
Exercise Scoring **Fully Trained (T)** = Satisfactory on 6 of 6 tasks **Need Practice (P)** = Satisfactory on 5 tasks **Untrained (U)** = Unsatisfactory on 4 or fewer tasks					

Table I-3. Mounted combat exercise.

During scoring procedures, the grader positions himself so that he can observe both the gunner and the target. Once the exercise commences, he:

- Times the tasks having a time standard.
- Observes and informs the gunner of the strike of the rounds.
- Records the results of each task in the right hand column with a "Y" for satisfactory completion of the task or an "N" for an unsatisfactory completion.
- Sums the results and assigns an overall score in accordance with the scores in the bottom box of the figure above.

d. **Night Fire Exercise** (Table I-4). This LFX develops and improves MK 19 gunnery skills at night or during degraded conditions. This exercise is conducted as often as the commander feels it is necessary to maintain crew skills.

ACTION	CONDITIONS	STANDARDS	AMMO	TIME	SAT Y/N
1. Mount the AN/TVS-5 to the MK 19.	At night, given an AN/TVS-5 complete, a MK 19, M3 tripod, and a T&E mechanism.	Gunner must mount the AN/TVS-5 on the MK 19 within 2 min.	0	2 min	
2. Engage a 400-meter BMP target using a zeroed MK 19 with an AN/TVS-5 mounted for observation.	At night, given an AN/TVS-5 complete, a MK 19, an M3 tripod, 6 rds of 40-mm TP, and a 400-meter BMP target, during the hours of daylight.	Gunner must impact 3 out of 6 rds on a BMP target at 400 meters within 1 min.	6 rds	1 min	
3. Engage a BMP target at 600 meters.	At night, given a MK 19, 8 rds of 40-mm TP, and a target at a distance of 600 meters.	Gunner must impact on a BMP target with at least 4 out of 8 rds at 600 meters within 2 min.	8 rds	2 min	
4. Engage a dismounted troop target at 800 meters.	At night, given a MK 19, 8 rds of 40-mm TP, and a target at a distance of 800 meters.	Gunner must impact within 5 meters of a silhouette target with at least 4 out of 8 rds at 800 meters within 2 min.	8 rds	2 min	
5. Dismount AN/TVS-5 from the MK 19.	At night, given a MK 19 and an AN/TVS-5.	Gun crew must take the AN/TVS-5 out of operation within 2 min.	0	2 min	
			Total: 22 rds		
Exercise Scoring **Fully Trained (T)** = Satisfactory on 5 of 5 tasks **Need Practice (P)** = Satisfactory on 4 tasks **Untrained (U)** = Unsatisfactory on 3 or fewer tasks					

Table I-4. Night fire exercise.

During scoring procedures, the grader positions himself so that he can observe both the gunner and the target. Once the exercise commences, he:

- Times the tasks having a time standard.
- Observes and informs the gunner of the strike of the rounds.
- Records the results of each task in the right hand column with a "Y" for a satisfactory completion of the task or a "N" for an unsatisfactory completion.

- Sums the results and assigns an overall score in accordance with the scores in the bottom box of the figure above.

I-2. COLLECTIVE GUNNERY

These three exercises test the application of fire from two or more guns against linear, deep, linear with depth, and area targets. They test collective individual skills, fire control, leader skills, adjustment of fire, methods of target engagement, and the control of one or more fire units. Tables I-5 through I-7 should be used with unit LFXs. They can be held on any range or piece of terrain that supports the MK 19. Targets may be made up of 55-gallon drums, hulks of old vehicles, silhouettes, chalk, panel markers, or engineer tape.

ACTION	CONDITIONS	STANDARDS	AMMO	TIME	SAT Y/N
1. Engage a linear target with a pair of MK 19s (two guns). (see also Task 2)	Given a linear target 100 meters wide from 800 to 1,200 meters, two MK 19s on tripods with T&E mechanisms, during the hours of daylight or under degraded conditions.	Gun crews must cover 90 percent of the linear target with effective fire while performing the following: 1. The leader lays the guns on his respective flanks. 2. The leader issues the fire command for engaging a linear target. 3. Gunners engage using traversing fire. 4. Observers ensure that fires remain on the target and adjust appropriately.	40 rds	2 min	
2. Engage a linear target with two pairs of MK 19s (four guns).	Given a linear target 200 meters wide from 800 to 1,200 meters, four MK 19s on tripods with T&E mechanisms, during the hours of daylight or under degraded conditions.	Same as Task 1, except the leader may have to subdivide the target.	80 rds	3 min	

Table I-5. Linear and deep targets.

ACTION	CONDITIONS	STANDARDS	AMMO	TIME	SAT Y/N
3. Engage a deep target with a pair of MK 19s (two guns).	Given a deep target 100 meters long with midrange to the target from 800 to 1,200 meters, two MK 19s on tripods with T&E mechanisms, during the hours of daylight or under degraded conditions.	Gun crews must cover 90 percent of the deep target with effective fire while performing the following: 1. The leader lays the guns on his respective ends of the target. 2. The leader issues the fire command for engaging a deep target. 3. Gunners engage using searching fire. 4. Observers ensure that fire remains on the target and adjust appropriately.	40 rds	2 min	
4. Engage a deep target with two pairs of MK 19s (four guns).	Given a deep target 200 meters long with midrange to the target from 800 to 1,200 meters, four MK 19s on tripods with T&E mechanisms, during the hours of daylight or under degraded conditions.	Gun crews should cover 90 percent of the deep target with effective fire while performing the same steps in task 3, except the leader may have to subdivide the target.	80 rds	3 min	
			Total: 240 rds		

Table I-5. Linear and deep targets (continued).

ACTION	CONDITIONS	STANDARDS	AMMO	TIME	SAT Y/N
1. Engage a linear target with depth using a pair of MK 19s (two guns).	Given a target 100 meters wide and 50 meters long (or 50 meters wide and 100 meters long), from 800 to 1,200 meters, two MK 19s with tripods and T&E mechanisms, during daylight or under degraded conditions.	Gun crews must cover 80 percent of the linear with depth target with effective fire while performing the following: 1. The leader lays the guns on the gun crews' respective flanks (ends) of the target. 2. The leader issues the fire command for engaging a linear target with depth. 3. Gunners engage using traversing and searching fire. 4. Observers ensure that fire remains on the target and adjust appropriately.	40 rds for each target.	3 min	
2. Engage a linear target with depth using two pairs of MK 19s (four guns).	Given a target 200 meters wide and 50 meters long (or 50 meters wide and 200 meters long), from 800 to 1,200 meters, four MK 19s with tripods and T&E mechanisms, during daylight or under degraded conditions.	Same as Task 1, except the leader may want to subdivide the target.	80 rds for each target.	4 min	
			Total: 120 rds		

Table I-6. Linear targets with depth.

ACTION	CONDITIONS	STANDARDS	AMMO	TIME	TPU
1. Engage an area target with two pairs of MK 19s (four guns).	Given an area target from 1,000 to 1,400 meters (midrange) extending over 200 meters in length and width, four MK 19s on tripods with T & E mechanisms, during the hours of daylight or under degraded conditions.	The combined gun crews must cover 70 percent of the area target with suppressive fire while performing the following: 1. The leader determines direction and elevation to the area target (midrange). 2. The leader divides the area target and ensures that each gun is laid on its portion, and assigns TRPs. 3. Each gun registers on its respective TRP. 4. The leader issues the fire command for an area target. 5. All MK 19s engage; gunners use traverse and search manipulations to adjust and distribute fires. 6. Observers ensure the fires remain on target and adjust appropriately.	160 rds	5 min	
2. Engage an area target with three pairs of MK 19s (six guns).	Given an area target from 1,000 to 1,400 meters (midrange), extending over 300 meters in length and 200 meters in width, six MK 19s on tripods with T&E mechanisms, during the hours of daylight or under degraded conditions.	The combined gun crews must cover 70 percent of the area target with suppressive fire while performing the following: (Performance measures are the same as for Task 1.)	240 rds	8 min	
			Total: 400 rds		

Table I-7. Area targets.

GLOSSARY

ACRONYMS AND ABBREVIATIONS

AAR	after-action review
AE	angle of elevation
AI	assistant instructor
AIT	advanced individual training
APC	armored personnel carrier
AS	angle of sight
BMP	fighting vehicle manufactured by the former Soviet Union
BOD	bore obstruction detector
BRDM	Soviet Union vehicle
C4	composition four
CALFEX	combined arms live-fire exercise
CLP	cleaner, lubricant, preservative
cm	centimeter
DA	Department of the Army
deg	degrees
DODIC	Department of Defense Identification Code
EA	engagement area
elev	elevation
EOD	explosive ordinance disposal
FM	field manual
FOV	field of view
fps	feet per second
ft	feet
FTX	field training exercise
GMD	grease, molybdenum disulfide
HE	high explosive
HEDP	high-explosive dual-purpose
HIMS	heavy interdiction missile system
HMMWV	high-mobility multipurpose wheeled vehicle
HWTS	heavy weapon thermal sight
IAW	in accordance with
IR	infrared
JP-4	jet propulsion fuel

LAW	lubricant, automatic weapons, arctic
LBE	load-bearing equipment
LED	light emitting diode
LFX	live-fire exercise
LH	left hand
LSA	lubricant oil, semifluid, automatic weapon
LSAT	lubricant oil, semifluid, automatic weapons, temperate
MACS	multipurpose arcade combat simulator
MGS	missile guidance system
min	minute
mm	millimeter
MOD	model
MOPP	mission-oriented protective posture
MPMG	multipurpose machine gun
MPRC	multipurpose range complex
MTP	mission training plan
MWTS	medium weapon thermal sight
NBC	nuclear, biological, chemical
NCO	noncommissioned officer
NCOES	noncommissioned officers education system
NFOV	narrow field of view
No.	number
NVD	night vision device
PEWS	platoon early warning system
PIBD	point initiating, base detonating
PMCS	preventive maintenance checks and services
QE	quadrant elevation
RBC	rifle bore cleaner
rds	rounds
RH	right hand
RP	reference point
RPG	rocket propelled grenade
sec	seconds
SOP	standing operating procedure
STX	situational training exercise
T&E	traverse and elevating
TESS	tactical engagement simulator system

TM	technical manual
TOE	table of organization and equipment
TP	training practice
TPIAL	target pointer illuminator aiming light
TPT	training practice trainer
TRADOC	U.S. Army Training and Doctrine Command
TRP	target reference point
TVPC	TOW vehicle power cable
TWS	thermal weapon sight
VDT	video disk trainer
W	width
WFOV	wide field of view

REFERENCES

SOURCES USED

These are the sources quoted or paraphrased in this publication.

DA Cir 350-38	Standards in Weapons Training. 3 July 1997.
DA Form 5517-R	Standard Range Cards. February 1986.
DA PAM 385-63	Range Safety. 19 May 2003.
FM 3-22.68	Crew-Served Machine Guns, 5.56-mm and 7.62-mm. 31 January 2003.
FM 3-25.26	Map Reading and Land Navigation. 20 July 2001.
FM 25-2	Unit Training Management. 10 September 1984.
FM 25-3	Training in Units. 12 December 1984.
TC 25-8/ with change	Training Ranges. 16 November 1992.

DOCUMENTS NEEDED

These documents must be available to the intended users of this publication.

* DA Form 5517-R	Standard Range Cards (LRA). February 1986.
DA Form 7518-R	MK 19 Day Practice and Qualification (Hull Targets).
DA Form 7519-R	MK 19 Night Practice and Qualification with Hull Targets and Night Vision Devices.
DA Form 7520-R	MK 19 Day Practice and Qualification (Pop-up Targets).
DA Form 7521-R	MK 19 Night Practice and Qualification with Pop-up Targets or without Night Vision Devices.
* TC 23-13	Crew-Served Weapon Night Vision Sight. 20 January 1967.

* This source was also used to develop this production.

* TM 9-1005-245-13&P/ with 5 changes	Operator's, Unit, and Direct Support Maintenance Manual with Repair Parts and Special Tools List for Machine Gun Mounts and Combinations for Tactical/Armored Vehicles and Ground Mounting M122A1 Machine Gun Tripod Mount, M3 Tripod Mount, M66 Machine Gun Mount Ring, M240E1 Flexible Machine Gun Mount, M142 Machine Gun Mount, .50 Caliber Machine Gun Mount. 22 March 2002.
* TM 9-1010-230-10	Operator's Manual for Machine Gun, 40-mm, MK 19, MOD 3. 30 May 2001.
* TM 9-1010-231-13&P	Operator's, Organizational and Intermediate Maintenance with Repair Parts List and Components List Mount, Machine Gun MK 64: MOD 4, 5, 6, 7, 8 used with Machine Gun, MK 19 MOD 3, 40-mm Machine Gun, M2, Caliber .50 Machine Gun, M60 Machine Gun, 7.62-mm Machine Gun, M60E3, 7.62-mm. 31 October 2000.
* TM 11-5855-214-10	Operator's Manual for Night Vision Sight, Crew-Served Weapon, AN/TVS-5. 15 February 1989.
* TM 11-5855-302-12&P	Operator's and Unit Maintenance Manual, Sight, Thermal, AN/PAS-13 (V2), AN/PAS-13 (V3). 1 January 1997.
* TM 11-5855-308-12&P	Operator's and Unit Maintenance Manual Target Pointer Illuminator/Aiming Light (TPIAL) AN/PEQ-2A. 15 May 2000.
* TM 11-5855-312-10	Operator's Manual, Sight, Thermal AN/PAS- 13B (V2), AN/PAS-13B (V3). 5 January 2001.
* TM 11-5860-201-10	Operator's Manual for Laser Infrared Observation Set, AN/GVS-5. 2 February 1982.

READINGS RECOMMENDED

These readings contain relevant supplemental information.

DoD Reg 5100-76-M	Physical Security of Sensitive Conventional Arms, Ammunition, and Explosives. August 2000.
FM 3-5/ with change	NBC Decontamination. 31 January 2002.
FM 3-21.71	Mechanized Infantry Platoon and Squad (Bradley). 20 August 2002.

* This source was also used to develop this production.

FM 5-250	Explosives and Demolitions. 30 July 1998.

FM 7-8/ with change	Infantry Rifle Platoon and Squad. 22 April 1992.
FM 7-20/ with change	The Infantry Battalion. 29 December 2000.
FM 31-70	Basic Cold Weather Manual. 12 April 1968.

INTERNET WEBSITES

Reimer Digital Library, http://www.adtdl.army.mil.

USAMC Logistics Support Activity, http://www.logsa.army.mil/.

Army Publishing Directorate, http://www.usapa.army.mil.

INDEX

MK 19, 40-mm GRENADE MACHINE GUN, MOD 3 FIRING TABLE I DAY PRACTICE AND QUALIFICATION WITH HULL TARGETS SCORECARD

For use of this form, see FM 3-22.27; the proponent agency is TRADOC.

PRIVACY ACT STATEMENT

AUTHORITY: 10 USC 3012(g)/Executive order 9397
PRINCIPAL PURPOSE: To aid individual training on targets at various ranges.
ROUTINE USES: To evaluate individual proficiency.
DISCLOSURE: Voluntary. However, mass rating and recording require some tracking method.

1a. LAST NAME	1b. FIRST NAME	1c. MI	2. RANK	3. UNIT

TABLE I (A). DISMOUNTED AND MOUNTED DAY PRACTICE

4. RANGE	5. LANE	6. GRADER	7. DATE

TASK	RANGE (Meters)	AMMO	TIME (Minutes)	ENGAGEMENT STANDARDS	GO	NO GO
ZERO	400	4	NA	2 ROUNDS HIT		
2	1,100 (+/- 200)	8	2.5	2 ROUNDS HIT		
3	1,500 (+/- 200)	10	3.5	2 ROUNDS HIT		
4	600 (+/- 100)	6	2	2 ROUNDS HIT		
5	800 (+/- 100)	6	2	2 ROUNDS HIT		
6	400	4	1.5	2 ROUNDS HIT		
MULTIPLE TARGETS						
7	1,100 (+/- 200)	10	4	1 ROUND HIT		
8	600 (+/- 100)			1 ROUND HIT		
9	800 (+/- 100)	14	4.5	1 ROUND HIT		
10	1,500 (+/- 200)			1 ROUND HIT		
				TOTALS		

8. COMMENTS

9. NUMBER OF ENGAGEMENT MET *(Choose One)*
- [ ] 10 - EXPERT
- [ ] 9 - SHARPSHOOTER
- [ ] 8-7 - MARKSMAN
- [ ] 6 AND BELOW - UNQUALIFIED

10. GUNNER'S SIGNATURE

11. GRADER'S SIGNATURE

TABLE I (B). DISMOUNTED AND MOUNTED DAY QUALIFICATION

12. RANGE	13. LANE	14. GRADER	15. DATE

TASK	RANGE (Meters)	AMMO	TIME (Minutes)	ENGAGEMENT STANDARDS	GO	NO GO
ZERO	400	4	NA	2 ROUNDS HIT		
2	1,100 (+/- 200)	8	2	2 ROUNDS HIT		
3	1,500 (+/- 200)	10	3	2 ROUNDS HIT		
4	600 (+/- 100)	6	1.5	2 ROUNDS HIT		
5	800 (+/- 100)	6	1.5	2 ROUNDS HIT		
6	400	4	1	2 ROUNDS HIT		
MULTIPLE TARGETS						
7	1,100 (+/- 200)	10	3.5	1 ROUND HIT		
8	600 (+/- 100)			1 ROUND HIT		
9	800 (+/- 100)	14	4	1 ROUND HIT		
10	1,500 (+/- 200)			1 ROUND HIT		
				TOTALS		

16. COMMENTS

17. NUMBER OF ENGAGEMENT MET *(Choose One)*
- [ ] 10 - EXPERT
- [ ] 9 - SHARPSHOOTER
- [ ] 8-7 - MARKSMAN
- [ ] 6 AND BELOW - UNQUALIFIED

18. GUNNER'S SIGNATURE

19. GRADER'S SIGNATURE

MK 19, 40-mm GRENADE MACHINE GUN, MOD 3 FIRING TABLE II
NIGHT PRACTICE AND QUALIFICATION WITH HULL TARGETS SCORECARD

For use of this form, see FM 3-22.27; the proponent agency is TRADOC.

PRIVACY ACT STATEMENT

AUTHORITY: 10 USC 3012(g)/Executive order 9397
PRINCIPAL PURPOSE: To aid individual training on targets at various ranges.
ROUTINE USES: To evaluate individual proficiency.
DISCLOSURE: Voluntary. However, mass rating and recording require some tracking method.

NOTE: Use these tables if the targets being used are hulls, and any of the following applies:
AN/PEQ-2A mounted on the TWS mounting bracket.
AN/PAS-13 mounted on the TWS mounting bracket.
AN/TVS-5 with the 3d generation tube mounted on the TWS mounting bracket.

1a. LAST NAME	1b. FIRST NAME	1c. MI	2. RANK	3. UNIT

TABLE II (A). DISMOUNTED AND MOUNTED NIGHT PRACTICE

4. RANGE	5. LANE	6. GRADER	7. DATE

TASK	RANGE (Meters)	AMMO	TIME (Minutes)	ENGAGEMENT STANDARDS	GO	NO GO
ZERO	400	4	NA	2 ROUNDS HIT		
2	1,100 (+/- 200)	8	2.5	2 ROUNDS HIT		
3	1,500 (+/- 200)	10	3.5	2 ROUNDS HIT		
4	600 (+/- 100)	6	2	2 ROUNDS HIT		
5	800 (+/- 100)	6	2	2 ROUNDS HIT		
6	400	4	1.5	2 ROUNDS HIT		
MULTIPLE TARGETS						
7	1,100 (+/- 200)	10	4	1 ROUND HIT		
8	600 (+/- 100)			1 ROUND HIT		
9	800 (+/- 100)	14	4.5	1 ROUND HIT		
10	1,500 (+/- 200)			1 ROUND HIT		
				TOTALS		

8. TYPE DEVICE *(Choose One)*
☐ AN/PEQ-2A ☐ AN/PAS-13 ☐ AN/TVS-5

9. NUMBER OF ENGAGEMENT MET *(Choose One)*
☐ 10 - EXPERT ☐ 8-7 - MARKSMAN
☐ 9 - SHARPSHOOTER ☐ 6 AND BELOW - UNQUALIFIED

10. GUNNER'S SIGNATURE	11. GRADER'S SIGNATURE

TABLE II (B). DISMOUNTED AND MOUNTED NIGHT QUALIFICATION

12. RANGE	13. LANE	14. GRADER	15. DATE

TASK	RANGE (Meters)	AMMO	TIME (Minutes)	ENGAGEMENT STANDARDS	GO	NO GO
ZERO	400	4	NA	2 ROUNDS HIT		
2	1,100 (+/- 200)	8	2	2 ROUNDS HIT		
3	1,500 (+/- 200)	10	3	2 ROUNDS HIT		
4	600 (+/- 100)	6	1.5	2 ROUNDS HIT		
5	800 (+/- 100)	6	1.5	2 ROUNDS HIT		
6	400	4	1	2 ROUNDS HIT		
MULTIPLE TARGETS						
7	1,100 (+/- 200)	10	3.5	1 ROUND HIT		
8	600 (+/- 100)			1 ROUND HIT		
9	800 (+/- 100)	14	4	1 ROUND HIT		
10	1,500 (+/- 200)			1 ROUND HIT		
				TOTALS		

16. TYPE DEVICE *(Choose One)*
☐ AN/PEQ-2A ☐ AN/PAS-13 ☐ AN/TVS-5

17. NUMBER OF ENGAGEMENT MET *(Choose One)*
☐ 10 - EXPERT ☐ 8-7 - MARKSMAN
☐ 9 - SHARPSHOOTER ☐ 6 AND BELOW - UNQUALIFIED

18. GUNNER'S SIGNATURE	19. GRADER'S SIGNATURE

MK 19, 40-mm GRENADE MACHINE GUN, MOD 3 FIRING TABLE III
DAY PRACTICE AND QUALIFICATION WITH POP-UP TARGETS SCORECARD

For use of this form, see FM 3-22.27; the proponent agency is TRADOC.

PRIVACY ACT STATEMENT

AUTHORITY: 10 USC 3012(g)/Executive order 9397

PRINCIPAL PURPOSE: To aid individual training on targets at various ranges.

ROUTINE USES: To evaluate individual proficiency.

DISCLOSURE: Voluntary. However, mass rating and recording require some tracking method.

1a. LAST NAME	1b. FIRST NAME	1c. MI	2. RANK	3. UNIT

TABLE III (A). DISMOUNTED AND MOUNTED DAY PRACTICE

4. RANGE	5. LANE	6. GRADER	7. DATE

TASK	RANGE (Meters)	AMMO	TIME (Minutes)	ENGAGEMENT STANDARDS	GO	NO GO
ZERO	400	4	NA	2 ROUNDS HIT		
2	600 (+/- 100)	6	2	1 ROUND HIT		
3	800 (+/- 100)	8	2.5	1 ROUND HIT		
4	400	4	1.5	1 ROUND HIT		
MULTIPLE TARGETS						
5	800 (+/- 100)	12	4	1 ROUND HIT		
6	400			1 ROUND HIT		
7	400	10	3	1 ROUND HIT		
8	600 (+/- 100)			1 ROUND HIT		
				TOTALS		

8. COMMENTS

9. NUMBER OF ENGAGEMENT MET *(Choose One)*

☐ 8 - EXPERT ☐ 6 - MARKSMAN

☐ 7 - SHARPSHOOTER ☐ 5 AND BELOW - UNQUALIFIED

10. GUNNER'S SIGNATURE

11. GRADER'S SIGNATURE

TABLE III (B). DISMOUNTED AND MOUNTED DAY QUALIFICATION

12. RANGE	13. LANE	14. GRADER	15. DATE

TASK	RANGE (Meters)	AMMO	TIME (Minutes)	ENGAGEMENT STANDARDS	GO	NO GO
ZERO	400	4	NA	2 ROUNDS HIT		
2	600 (+/- 100)	6	1.5	1 ROUND HIT		
3	800 (+/- 100)	8	2	1 ROUND HIT		
4	400	4	1	1 ROUND HIT		
MULTIPLE TARGETS						
5	800 (+/- 100)	12	3.5	1 ROUND HIT		
6	400			1 ROUND HIT		
7	400	10	2.5	1 ROUND HIT		
8	600 (+/- 100)			1 ROUND HIT		
				TOTALS		

16. COMMENTS

17. NUMBER OF ENGAGEMENT MET *(Choose One)*

☐ 8 - EXPERT ☐ 6 - MARKSMAN

☐ 7 - SHARPSHOOTER ☐ 5 AND BELOW - UNQUALIFIED

18. GUNNER'S SIGNATURE

19. GRADER'S SIGNATURE

MK 19, 40-mm GRENADE MACHINE GUN, MOD 3 FIRING TABLE IV
NIGHT PRACTICE AND QUALIFICATION WITH POP-UP TARGETS SCORECARD

For use of this form, see FM 3-22.27; the proponent agency is TRADOC.

PRIVACY ACT STATEMENT

AUTHORITY: 10 USC 3012(g)/Executive order 9397
PRINCIPAL PURPOSE: To aid individual training on targets at various ranges.
ROUTINE USES: To evaluate individual proficiency.
DISCLOSURE: Voluntary. However, mass rating and recording require some tracking method.

NOTE: Use this table if you do not have a MK 19/sight combination that applies to Table II, and you are using hull targets OR if targets are pop-up/E type silhouettes.

1a. LAST NAME	1b. FIRST NAME	1c. MI	2. RANK	3. UNIT

TABLE IV (A). DISMOUNTED AND MOUNTED NIGHT PRACTICE

4. RANGE	5. LANE	6. GRADER	7. DATE

TASK	RANGE (Meters)	AMMO	TIME (Minutes)	ENGAGEMENT STANDARDS	GO	NO GO
ZERO	400	4	NA	2 ROUNDS HIT		
2	600 (+/- 100)	6	2	2 ROUNDS HIT		
3	800 (+/- 100)	6	2	2 ROUNDS HIT		
4	400	4	1.5	2 ROUNDS HIT		
MULTIPLE TARGETS						
5	800 (+/- 100)	10	3.5	1 ROUND HIT		
6	400			1 ROUND HIT		
7	400	14	2.5	1 ROUND HIT		
8	600 (+/- 100)			1 ROUND HIT		
				TOTALS		

8. COMMENTS

9. NUMBER OF ENGAGEMENT MET *(Choose One)*

☐ 8 - EXPERT ☐ 6 - MARKSMAN
☐ 7 - SHARPSHOOTER ☐ 5 AND BELOW - UNQUALIFIED

10. GUNNER'S SIGNATURE	11. GRADER'S SIGNATURE

TABLE IV (B). DISMOUNTED AND MOUNTED NIGHT QUALIFICATION

12. RANGE	13. LANE	14. GRADER	15. DATE

TASK	RANGE (Meters)	AMMO	TIME (Minutes)	ENGAGEMENT STANDARDS	GO	NO GO
ZERO	400	4	NA	2 ROUNDS HIT		
2	600 (+/- 100)	6	1.5	2 ROUNDS HIT		
3	800 (+/- 100)	6	1.5	2 ROUNDS HIT		
4	400	4	1	2 ROUNDS HIT		
MULTIPLE TARGETS						
5	800 (+/- 100)	10	3	1 ROUND HIT		
6	400			1 ROUND HIT		
7	400	14	2	1 ROUND HIT		
8	600 (+/- 100)			1 ROUND HIT		
				TOTALS		

16. COMMENTS

17. NUMBER OF ENGAGEMENT MET *(Choose One)*

☐ 8 - EXPERT ☐ 6 - MARKSMAN
☐ 7 - SHARPSHOOTER ☐ 5 AND BELOW - UNQUALIFIED

18. GUNNER'S SIGNATURE	19. GRADER'S SIGNATURE

DA FORM 7521-R, AUG 2003 APD V1.00

Made in the USA
Monee, IL
05 November 2026

67980406R00177